VOLUME EIGHTY SEVEN

CURRENT TOPICS IN DEVELOPMENTAL BIOLOGY

Development of Neural Circuitry

VOLUME EIGHTY SEVEN

Current Topics in Developmental Biology

Development of Neural Circuitry

Edited by

OLIVER HOBERT
Department of Biochemistry
Columbia University
New York, USA

ELSEVIER

AMSTERDAM • BOSTON • HEIDELBERG • LONDON
NEW YORK • OXFORD • PARIS • SAN DIEGO
SAN FRANCISCO • SINGAPORE • SYDNEY • TOKYO
Academic Press is an imprint of Elsevier

Academic Press is an imprint of Elsevier
525 B Street, Suite 1900, San Diego, CA 92101-4495, USA
30 Corporate Drive, Suite 400, Burlington, MA 01803, USA
32, Jamestown Road, London NW1 7BY, UK
Linacre House, Jordan Hill, Oxford OX2 8DP, UK

First edition 2009

ISBN: 978-0-12-374469-2
ISSN: 0070-2153

For information on all Academic Press publications
visit our website at elsevierdirect.com

Printed and bound in USA
09 10 11 9 8 7 6 5 4 3 2 1

Contents

Contributors

Claire Bénard
Department of Biochemistry and Molecular Biophysics, Howard Hughes Medical Institute, Columbia University Medical Center, New York, USA

Renata Batista-Brito
Smilow Neuroscience Program, and the Department of Cell Biology, New York University Medical Center, New York, USA and the Gulbenkian PhD Programme in Biomedicine, Gulbenkian Science Institute, Oeiras, Portugal

Daniel A. Colón-Ramos
Program in Cellular Neuroscience, Neurodegeneration and Repair, Department of Cell Biology, Yale University School of Medicine, New Haven, Connecticut, USA

Jeremy S. Dasen
Smilow Neuroscience Program, Department of Physiology and Neuroscience, New York University School of Medicine, New York, New York, USA

Gord Fishell
Smilow Neuroscience Program, and the Department of Cell Biology, New York University Medical Center, New York, USA

Fred H. Gage
Laboratory of Genetics LOG-G, The Salk Institute for Biological Studies, La Jolla, California, USA

Geoffrey J. Goodhill
Queensland Brain Institute and School of Mathematics and Physics, The University of Queensland, St Lucia, Australia

Oliver Hobert
Department of Biochemistry and Molecular Biophysics, Howard Hughes Medical Institute, Columbia University Medical Center, New York, USA

Yan Li
Laboratory of Genetics LOG-G, The Salk Institute for Biological Studies, La Jolla, California, USA

Duncan Mortimer
Queensland Brain Institute, The University of Queensland, St Lucia, Australia

Yangling Mu
Laboratory of Genetics LOG-G, The Salk Institute for Biological Studies, La Jolla, California, USA

Hugh D. Simpson
Queensland Brain Institute, The University of Queensland, St Lucia, Australia

Preface

With its multiplicity of cell types and complex patterns of cellular interactions, the nervous system represents the most complex organ of a metazoan organism. Understanding how neuronal circuits develop is one of the holy grails of developmental biology. This volume sets out to review some selected areas of research in which some progress has been made over the past few years. A range of model organisms from worms to vertebrates is covered, and experimental as well as theoretical aspects of this problem are discussed. I thank the contributors to this volume and the editorial staff at Elsevier for making the volume happen.

Oliver Hobert

CHAPTER ONE

Theoretical Models of Neural Circuit Development

Hugh D. Simpson,* Duncan Mortimer,* *and* Geoffrey J. Goodhill*,†

Contents

Abstract

Proper wiring up of the nervous system is critical to the development of organisms capable of complex and adaptable behaviors. Besides the many experimental advances in determining the cellular and molecular machinery that carries out this remarkable task precisely and robustly, theoretical approaches have also proven

* Queensland Brain Institute, The University of Queensland, St Lucia, Australia
† School of Mathematics and Physics, The University of Queensland, St Lucia, Australia

Current Topics in Developmental Biology, Volume 87
ISSN 0070-2153, DOI: 10.1016/S0070-2153(09)01201-0

to be useful tools in analyzing this machinery. A quantitative understanding of these processes can allow us to make predictions, test hypotheses, and appraise established concepts in a new light. Three areas that have been fruitful in this regard are axon guidance, retinotectal mapping, and activity-dependent development. This chapter reviews some of the contributions made by mathematical modeling in these areas, illustrated by important examples of models in each section. For axon guidance, we discuss models of how growth cones respond to their environment, and how this environment can place constraints on growth cone behavior. Retinotectal mapping looks at computational models for how topography can be generated in populations of neurons based on molecular gradients and other mechanisms such as competition. In activity-dependent development, we discuss theoretical approaches largely based on Hebbian synaptic plasticity rules, and how they can generate maps in the visual cortex very similar to those seen *in vivo*. We show how theoretical approaches have substantially contributed to the advancement of developmental neuroscience, and discuss future directions for mathematical modeling in the field.

1. Introduction

Wiring up the developing nervous system involves a vast array of challenges. Billions of connections must be made appropriately, but rather than uniquely specifying each of these connections, nervous systems appear to adopt strategies for development based on general rules which apply across large numbers and patterns of connections. Such general rules include, for instance, how to build a topographic map based on molecular gradients and neural firing patterns, and how to change the strengths of synapses based on correlations in presynaptic and postsynaptic firing. As discussed in the other chapters of this volume, well-designed experimental approaches, combined with rapidly developing technologies for imaging the developing nervous system, have dramatically expanded our understanding of some of these general strategies, and their specific molecular implementations. However, in common with most of biology, this understanding is mostly qualitative. While the "thinking models" developed are extremely useful, they are usually not up to the task of making fully *quantitative* predictions about outcomes in different situations. Qualitative models can rapidly become overwhelmed by the huge complexity of the data, whereas quantitative models offer the potential to tease apart which variables and interactions are key to understanding the phenomena under consideration (for discussions, see Abbott, 2008; Goodhill, 2007; Mogilner *et al.*, 2006).

To illustrate these general issues, it is interesting to draw an analogy with the historical progression of our understanding of certain physical systems. Consider, for example, the effect of pressure and volume on the temperature of gases. It is relatively straightforward to obtain qualitative data demonstrating the general relationships between these variables. More quantitative measurements are more challenging, but begin to suggest lawful relationships between the variables. But it is only when these relationships are expressed in the form of a mathematical equation (the Ideal Gas Law) that one has the power to predict behavior under a wide variety of circumstances. Furthermore, the equation provides a target for understanding *why* these relationships exist. Indeed, subsequent theoretical work showed that the Ideal Gas Law could be derived from a statistical–mechanical approach, basing it firmly in the behavior of individual molecules. However, while the Ideal Gas Law is fundamental to our understanding of gases, it ignores key effects such as intermolecular interactions, and therefore fails under certain circumstances. Thus, mathematical descriptions can still be extremely powerful tools, even when they clearly fail to include details of the real world which are known to be important. In fact, essentially all mathematical models of physical phenomena ignore details which may be important under some circumstances. It is thus not surprising that analogous limitations in scope are true of mathematical models in biology. As demonstrated by the Ideal Gas Law, this limited scope does not mean these models are not insightful and useful.

Of course, even the most optimistic computational neuroscientist would not claim that theoretical modeling of neural wiring development has so far resulted in anything like the physical laws mentioned above. Nonetheless, there are several domains where mathematical models have clarified our understanding of the capabilities and limitations of potential developmental mechanisms (van Ooyen, 2003). In this chapter, we will focus on three representative examples: axon guidance, retinotectal map formation, and activity-dependent development. In the first section, we look at the mathematical approaches that have been applied to understanding a variety of phenomena in growth cone behavior. For the most part, specific unifying principles have not emerged from these models; hence we organize models by general approach taken. In retinotectal map formation theoretical models have a long history of close engagement with experimental data, ranging from older surgically induced map manipulations, to more recent data regarding genetic manipulations of Eph and ephrin gradients. In this section, we organize models historically. In activity-dependent development, the dominant paradigm has been to apply mathematical models of Hebbian learning to a range of developmental phenomena, for instance in the visual system. In this section, we therefore start by focusing on one of the simplest mathematical models of Hebbian learning, and then discuss more sophisticated approaches based on this and other principles. Given that the rest of this volume reviews experimental data in some detail, here we assume a basic knowledge of this data and focus instead on the mathematical models.

2. Theoretical Modeling in Axon Guidance

To form the precise arrangement of connections between neurons seen in mature nervous systems, axons must be accurately directed to find their targets. This process depends heavily on the interaction between the growth cone and its environment. Current modeling of growth cone guidance and dynamics is a fragmented field. This is largely due to a dearth of quantitative data on growth cone behavior, which makes it difficult to unify experimental findings under one framework. Thus, to survey the relevant literature, we divide models into three rough categories based on the modeling approach taken: phenomenological, mechanistic, and abstract (see Maskery and Shinbrot, 2005 for another review).

2.1. Phenomenological models

Phenomenological models describe experimentally observed behavior by fitting it to a mathematical framework without making explicit reference to underlying mechanisms; much like Boyle's empirical observations of the inverse relationship between pressure and volume of a gas (though alas, seldom as mathematically concise for growth cones). This style of model has been applied to microtubule dynamics (Odde and Buettner, 1998; Odde *et al.*, 1996), filopodial dynamics (Buettner, 1995; Odde and Buettner, 1998), the "random walk" behavior of the growth cone (Katz *et al.*, 1984), and interactions between filopodia and target cells (Buettner, 1996).

Buettner's work (Buettner, 1995; Buettner *et al.*, 1994) provides a good example of this kind of model. By statistically analyzing time-lapse images of growth cones undergoing dynamic changes in morphology (Buettner *et al.*, 1994), Buettner was able to develop some probabilistic rules regarding filopodial dynamics, which were formalized as a model in Buettner (1995). Growth cone morphology is described by the instantaneous length and angle of each filopodium, and the dynamics of the filopodia are characterized by five parameters: the rate at which filopodia extend, the rate of retraction, the average rate at which new filopodia are initiated (modeled as a Poisson process), and two parameters for the scale and shape of a gamma-distribution for the time over which a filopodium extends before it begins to retract. Along with a simple probabilistic rule for where a filopodium initiates, and under the assumption that filopodia extend radially from the center of the growth cone, this model gives qualitatively realistic morphologies, which also satisfied quantitative constraints such as the correct average number of filopodia. Buettner then analyzed model growth cones with different parameter values, simulating their interaction with a contact-mediated guidance cue at different distances from the growth cone body. From this, she was able to extract some useful quantities; for example, the ratio of extension and retraction rates, which determines the mean time required before the first filopodium on the growth cone contacts the cue. By mapping the effects of external cues such as guidance molecules onto the parameters of the model, one can hope to gain some intuition as to how those cues might operate.

Growth cone behavior and axon extension are thought to be mediated by partially independent, but related, processes. This has led to modeling studies focused on characterizing the interaction between the two. Work in the mid-1980s (Katz *et al.*, 1984) argued that under some circumstances axon elongation can be regarded as a one-dimensional uncorrelated random walk, with parameters varying with neuronal type, substrate, and chemical environment. In the mid-1990s, further work in this vein demonstrated that, rather than being uncorrelated, axon extension exhibits significant anticorrelation on short timescales—4 min, to be precise (Odde *et al.*, 1996). In other words, there is a tendency for any increases in axonal length to be followed by a retraction with a lag of about 4 min. In the same article, the authors also showed significant correlation between the dynamics of microtubule polymerization and growth cone advance (Fig. 1.1).

Such studies, while not in themselves providing explanations for why the observed behaviors occur, are nonetheless very useful for the precise constraints they suggest for growth cone behavior, and the hypotheses they

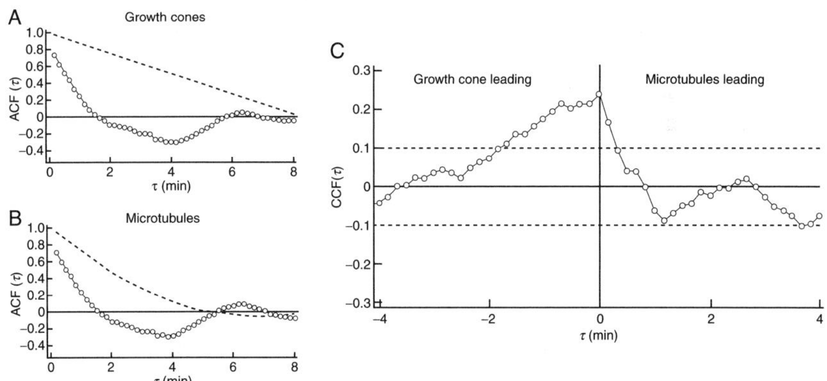

Figure 1.1 Temporal correlation in microtubule polymerization and growth cone extension. The detrended position autocorrelation functions (ACF) for growth cone and microtubule advance ((A) and (B) respectively) are plotted against time (unfilled circles). For comparison, the dashed line shows the autocorrelation function for a random walk generated by randomly shuffling measured displacements from the original data series. The observed form of the true autocorrelation function illustrates that net growth cone or microtubule advance tends to be followed by retraction about 4 min later. (C) Plotting the cross-correlation function (CCF) between growth cone and microtubule dynamics illustrates strong interactions between the two processes (from Odde *et al.*, 1996; reproduced with permission from John Wiley & Sons, Inc.).

spark concerning how underlying molecular events might contribute to the observed dynamics. For example, obtaining a better quantitative understanding of the connection between microtubule polymerization and growth cone advance through crosscorrelation analysis may aid in understanding the role these cytoskeletal elements play in growth cone motility.

2.2. Mechanistic models

Mechanistic models focus on actual biophysical mechanisms and their role in growth cone behavior; for example, the dynamics of certain cytoskeletal components and how they affect the rate of neurite growth (Hely and Willshaw, 1998; Kiddie *et al.*, 2005; Mogilner and Rubinstein, 2005), the spatiotemporal calcium concentration distributions underlying growth cone turning (Aeschlimann and Tettoni, 2001), or the patterns of activity of the Rho GTPases and their influence on growth cone guidance (Sakumura *et al.*, 2005). Such models are difficult to construct at present, partly due to the complexity of growth cone biochemistry, and partly due to the lack of experimental data on important quantities such as reaction rate constants, as well as concentrations of and interactions between molecular species.

Experimental evidence suggests that tubulin molecules are synthesized almost exclusively in the soma, and then assembled into microtubules predominantly in the growth cone (Kobayashi and Mundel, 1998). This implies that axon outgrowth is limited by the rate at which microtubules can be transported to regions of active extension. A number of theoretical models have explored this idea, and included various effects such as diffusive and active transport of tubulin monomers, competition between neurites for tubulin, viscoelastic stretching of axon segments, calcium-induced microtubule depolymerization, and varying intrinsic rates of tubulin polymerization and depolymerization within different growth cones (Graham and van Ooyen, 2001; Kiddie *et al.*, 2005; McLean *et al.*, 2004; van Veen and van Pelt, 1994; reviewed in van Ooyen, 2001). These models have been successively refined, ultimately incorporating compartment-based modeling with dynamic compartment allocation (Graham and van Ooyen, 2001; Kiddie *et al.*, 2005). Most strikingly, this modeling program has demonstrated that small variations in polymerization and depolymerization rates in the growth cones of different neurites can lead to sharp changes in elongation rate, resulting in growth cone pausing and neurite retraction.

Mogilner and Rubinstein's (2005) exploration of the physics underlying filopodial extension is an excellent, and unusually successful example of this form of modeling, applicable to many systems besides growth cones. Length, growth rate, and spacing between adjacent filopodia are analyzed by modeling the effects of membrane elasticity, g-actin diffusion, lamellipodial extension rate, and the buckling of f-actin bundles under strain. This analysis indicates that the length of a filopodium is determined by the number of bundled actin filaments in its core. For less than about 10 bundled actin filaments, the strain exerted on the bundle by the membrane is sufficient to cause buckling for very short filopodia. As the number of included filaments increases, it becomes less likely for the filopodium to buckle, but at the same time more g-actin is required for the structure to continue extending, so that when the number of filaments is too large, the filopodium is also unable to extend. This led the authors to calculate an optimal value of approximately 30 actin filaments, giving the best trade-off between mechanical stability and g-actin diffusion (Fig. 1.2). In conclusion, the authors obtained average lengths for extending filopodia of between 1 and 10 μm, values which agree well with experiment.

This model is, however, far from being able to predict the responses of growth cones to a guidance cue gradient, and it is this task that the model of Sakumura *et al.* (2005) tackles. This model is based on the well known and widely studied Rho GTPase system. The Rho GTPases are a family of proteins known to play important roles in actin-based motility in a number of cellular systems (Giniger, 2002). The authors recognize the lack of information about the kinetic parameters characterizing the interactions

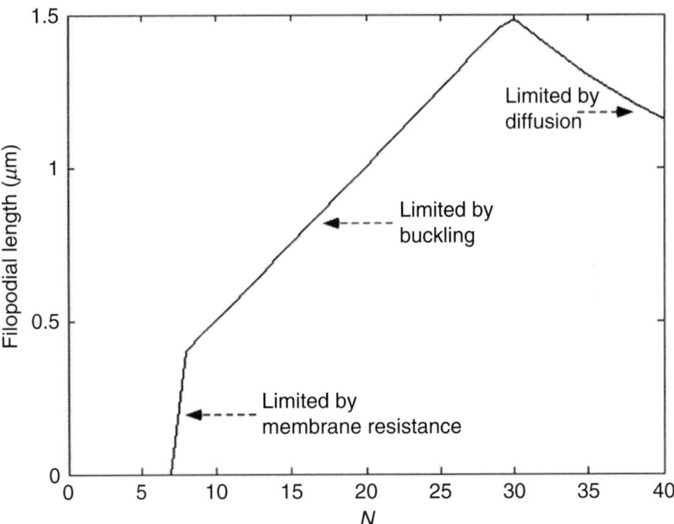

Figure 1.2 Filopodial length is limited by diffusion and buckling. Depending on the number of bundled actin fibers, N, various effects act to limit the length of filopodia. For too few bundled fibers, the actin polymerization force is not sufficient to overcome membrane resistance, and filopodial growth is strongly attenuated. For more than about eight bundled fibers, filopodia can overcome membrane resistance, but buckle when they reach a length dependent on the number of fibers. Finally, for greater than about 30 fibers, diffusion cannot deliver unpolymerized g-actin to the tips of the filopodia fast enough to maintain growth, and extension stalls (from Mogilner and Rubinstein, 2005; reproduced with permission from Elsevier).

between the Rho GTPases, and instead take a qualitative approach, randomly assigning values to various unknown parameters, then classifying the behavior of the dynamical system which results. The main conclusion drawn from this analysis is that for a large range of kinetic parameters, the Rho GTPase network exhibits hypersensitivity to changes in concentration of a particular class of molecules, the Rho guanosine exchange factors (Rho GEFs). As the concentration of Cdc42-GEFs increases above a threshold, the system dynamics moves from stable attractor into a rapid oscillatory regime (Fig. 1.3). The authors posit that this oscillation is characteristic of the streamlined mode of growth cone movement, while the stable attractor corresponds to the "slow-expansion" mode in which the growth cone takes on a more complex morphology. However, a more recent analysis of the Rho GTPase network, assuming a slightly different network connectivity, finds multistable rather than oscillatory solutions (Jilkine *et al.*, 2007).

At this stage, fully mechanistic models are underconstrained due to a lack of quantitative data for crucial parameters. Ideally, such models require knowledge of reaction rate constants, concentrations of key molecules inside the growth cone and an accurate map of interactions between the species of molecules present. Finally, it is worth noting that axon guidance shares many similarities with eukaryotic chemotaxis (Mortimer *et al.*, 2008; von Philipsborn and Bastmeyer, 2007), a field for which mathematical models have proved very useful.

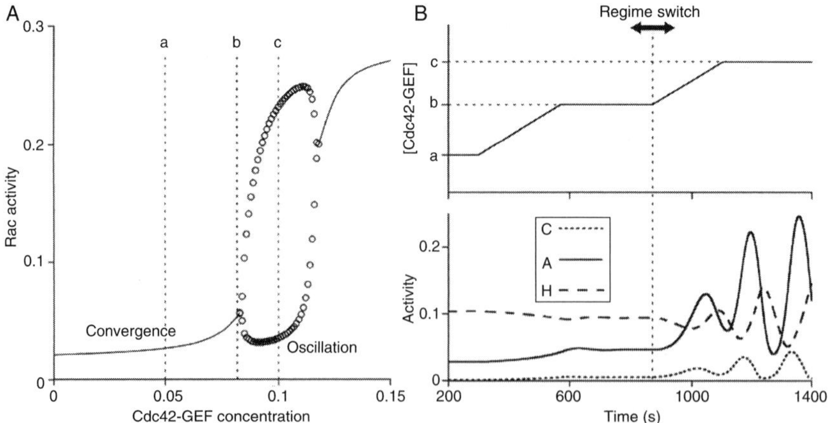

Figure 1.3 Rho GTPase network dynamics can depend sensitively on Cdc42-GEF concentration. (A) For sufficiently low concentrations of Cdc42-GEF, the network settles into a stable state. As Cdc42-GEF concentration is increased, a threshold is passed at which the network enters a large-amplitude oscillatory mode. (B) External control of Cdc42-GEF concentration (perhaps by receptor activation) can induce a regime change in the activation dynamics of Cdc42 (dashed line), Rac (solid line), and RhoA (dotted line) (from Sakumura *et al.*, 2005; reproduced with permission from Elsevier).

2.3. Abstract models

Abstract models discuss the mechanism(s) underlying a particular behavior in more general terms, without explicit reference to the molecular details of its implementation. The influential model of Meinhardt (1999) provides an archetypal example of this kind of approach; it describes a general strategy for the growth cone to amplify weak signals from its environment. In Meinhardt's model, this amplification is achieved by coupling the external signal to a pattern formation system involving local activation and long range inhibition. The system begins in a spatially symmetric, but unstable steady state. Symmetry is broken by the external signal, which pushes the system into a stable, asymmetric state reflecting the direction in which the symmetry was broken. One difficulty with this approach, recognized by Meinhardt, is that the system then becomes stuck: it is unable to respond to new inputs, such as a change in the external signal. Meinhardt works around this by postulating a second mechanism which serves to reset the system to its original, unstable state. He also makes several other generalizations of his basic model which qualitatively capture other aspects of chemotactic devices, such as the formation of filopodia.

The work of Goodhill, Urbach, and Baier in the late 1990s (Goodhill, 1997; Goodhill and Baier, 1998; Goodhill and Urbach, 1999; Urbach and Goodhill, 1999) following Berg and Purcell (1977) can also be placed in this category. Growth cones are believed to sense and respond to gradients by comparing receptor binding across their spatial extent: the side of a growth cone exposed to the highest concentration of ligand will, on average, also display the largest amount of receptor binding. If growth cones do use such a spatial-sensing strategy (cf. temporal sensing for bacteria), then in order for a growth cone to detect and reliably respond to a chemical gradient, the noise due to fluctuations in receptor binding cannot be much larger than the difference in receptor binding across its spatial extent. By modeling the physics of receptor–ligand interaction, Goodhill and Urbach estimated the limitations growth cones face when responding to chemical gradients. If the root mean-squared error in a concentration measurement is given by σ_C, then the error associated with taking the difference between two such measurements is $\sigma_{\Delta C} = \sqrt{2}\sigma_C$. This gives an order-of-magnitude lower bound on the difference in concentration, ΔC_{min}, that the growth cone can detect: $\Delta C_{\mathrm{min}} \approx \sqrt{2}\sigma_C$. A strong point of this approach is that the results should apply regardless of the intracellular mechanism used to extract the gradient direction from the spatiotemporal distribution of bound receptors—the details of the intracellular machinery can only reduce performance. Hence, their results give an upper bound on gradient-sensing performance of growth cones.

Abstract models have also been developed which simulate growth cone behavior for direct comparison to experiment, rather than discussing general restrictions on gradient sensing. Goodhill *et al.* (2004) examined the

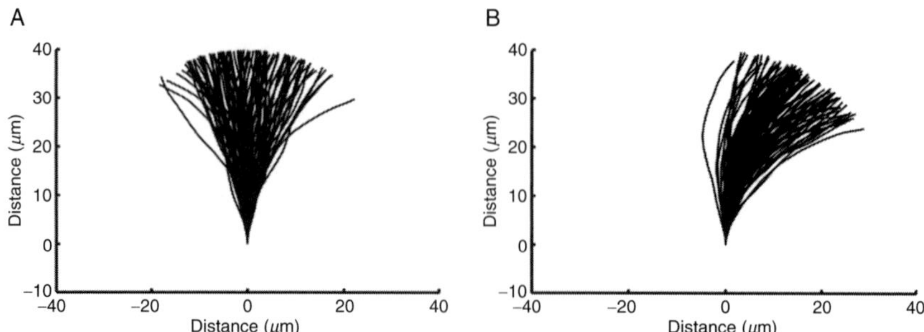

Figure 1.4 Growth cone trajectories generated by a filopodium-based model. (A) Without exposure to a guidance cue, axons project straight on average. (B) In contrast, when exposed to a 5% gradient pointing to the right, the axons turn (from Goodhill *et al.*, 2004).

implications of different models for filopodial formation and the subsequent influence of filopodia on the growth cone (Fig. 1.4). One finding from this work was that, given some simple assumptions about how an external signal influences the distribution of filopodia on the growth cone, testable experimental predictions could be made—namely, that the growth cone may display qualitatively different sensitivity curves to attractive and repulsive gradients. A further example is provided by Xu *et al.* (2005) who forgo any speculation about the role of specific structures in growth cone chemotaxis, focusing instead on the issues of adaptation, temporal averaging, and spatial averaging. These ideas are then applied to a simplified model growth cone which nonetheless displays realistic trajectories when appropriate parameters are chosen. This work demonstrated that experimentally observed behaviors displayed when growth cones are exposed to a gradient for a long period of time (Rosoff *et al.*, 2004), can be reproduced without explicitly implementing an adaptation mechanism. This suggests that experimental work probing growth cone behavior on short timescales (Ming *et al.*, 2002; Piper *et al.*, 2005) may require reinterpretation (e.g., see Mortimer *et al.*, 2008).

Such direct simulation models have also been applied to understanding the kinds of effects that come into play when more than one axon is considered. One of the earliest examples of computer modeling in axon guidance (Katz and Lasek, 1985) studied the formation of axon sheets and axon bundles in the vertebrate spinal cord. In this approach, axon extension was modeled as a simple random walk on a hexagonal lattice. Three additional constraints were investigated to determine which were necessary to produce ordered sheets of axon extension: directional persistence (achieved by limiting the degree to which an individual axon could bend), differential adhesion (in which axons preferred binding to the substrate than to other axons), and initial polarization (in which axons were assumed to initially project in a uniform direction). By systematically investigating the

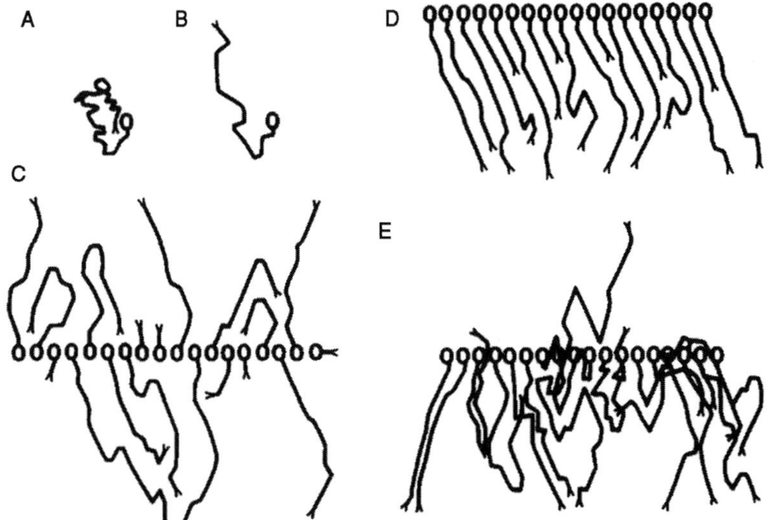

Figure 1.5 Models can be used to test the importance of various assumptions. In the model of Katz and Lasek (1985), the degree to which axons wander can vary (compare (A) with (B)); neurons can project in an unoriented or oriented manner (compare (C) with (D)), and axon crossing can be differentially inhibited (compare (D) with (E)). Only in case (D) do the axons project in ordered sheets (from Katz and Lasek, 1985; reproduced with permission from Elsevier).

effect of strengthening or weakening these constraints, the authors found necessary conditions to ensure the growth of ordered sheets of axons; in particular, that for a dense enough field of axons, the inhibition of axon crossing along with initial polarization of outgrowth direction was sufficient for ordering (Fig. 1.5). Hentschel and van Ooyen (1999) and Krottje and van Ooyen (2007) are more recent simulation models that allow for interactions between multiple axons. The latter provides a general framework for studying systems level effects in axon guidance. Here, the authors develop a set of mathematical tools allowing the simulation of the development of multiple axons through complex domains including any number of sources of guidance factors (which can even be released by the growth cones themselves).

3. Theoretical Modeling in the Retinotectal System

The ordered projection from the retina to the midbrain tectum is often referred to as the retinotectal map. (The corresponding midbrain target in mammals is the superior colliculus; hence also retinocollicular map.)

The map formed by these connections is described as topographic or retinotopic because neighborhood spatial relationships are preserved; that is, two cells that are near each other in the retina project to neighboring positions in the tectum (Fig. 1.6).

The development of midbrain retinotopy requires the axons of spatially ordered cells in the retina to navigate through a complex three-dimensional environment to their target, and re-establish their original order. We focus

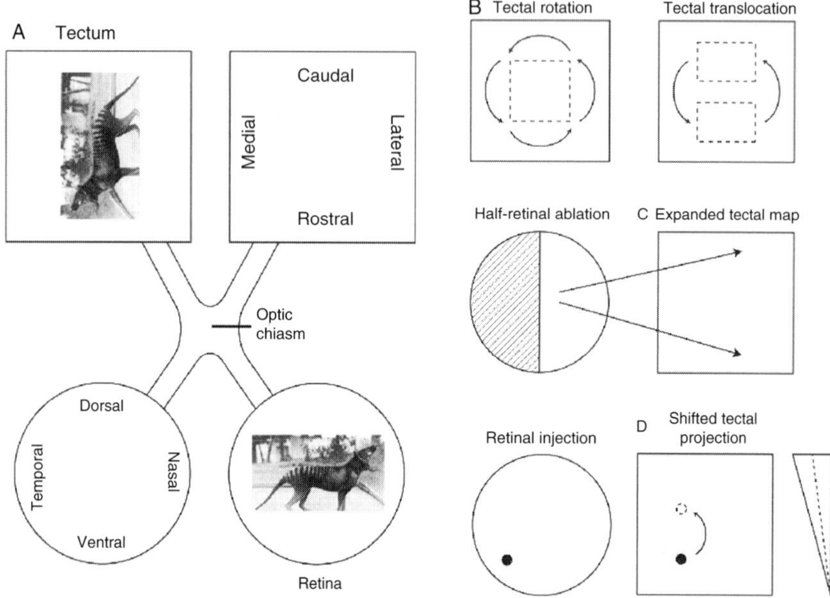

Figure 1.6 Normal and experimental variants in retinotectal maps. (A) Schematic of the normal retinotectal projection; nasal (temporal) retina maps to caudal (rostral) tectum and dorsal (ventral) retina maps to lateral (medial) tectum. In this way, an image (here the presumed extinct Tasmanian Tiger) presented to the retina is reproduced in the pattern of activity in the midbrain. (B) Schematics of tectal graft manipulations. On the left, a central piece of tectum is rotated (along with any markers on the tectum); maps formed after these procedures were also rotated. On the right, a graft translocation experiment is pictured, where two pieces of tectum have their positions exchanged. In this case, the corresponding map is often found to have similar translocations. (C) Retinal ablation experiment. Half of the retina is ablated, and instead of remaining confined to its usual half-tectum, the projection of the nonablated retina axons expands to fill the entire tectum. (D) Schematic (simplified) of a more recent misexpression study. Here, a gradient of repulsive guidance molecules (solid triangle, right) is distributed along the rostrocaudal axis of the tectum. Normally, this gradient is involved in determining where retinal ganglion cell axons project to. Here, for example, a group of ventrotemporal cells in the retina (filled circle, left) project to rostromedial tectum (filled circle, center). If this gradient is artificially weakened (dotted line, right), the reduced level of repulsion allows axons from the same retinal origin to terminate more caudally (dashed circle, center).

on the final step in this sequence: how the axonal terminations order themselves correctly on the tectum. Mathematical models of this process generally fall into the abstract category described in Section 2, and so are better categorized as belonging to one of the two main generations of modeling (and experiments) so far (1) earlier systems-based data and models and (2) more recent molecular level data and models. The large number of detailed models built around the systems results necessitates more brief descriptions, while the more recent models are considered in greater detail. This should not be taken as a suggestion that earlier models have been uniformly improved on by newer models that utilize the latest data on the retinotectal system. In fact earlier models tended to be more flexible and mathematically sophisticated than more recent ones, so that the core problems in retinotectal modeling will likely require insights from both generations. We preface the discussion of these generations of models with a brief description of hypotheses relevant to retinotectal mapping.

3.1. Hypotheses underlying map formation

A number of mechanisms have been identified that make significant contributions to retinotectal map formation. Models generally employ at least one of the following:

1. The *chemoaffinity* hypothesis of retinotectal mapping posits that gradients of molecular markers across a population of cells allow positions of cells to be uniquely identified by the level of marker on the cell, and that if there is a second population of cells with matched gradients, then this can allow for a topographic map to form between them (Sperry, 1963). Such gradients can also provide vector signals to guide ingrowing axons.
2. *Competition*, usually between retinal ganglion cell (RGC) axons for some limiting resource such as target space, has been consistently used in models from both generations (Fraser and Perkel, 1990; Gaze and Keating, 1972; Goodhill and Xu, 2005; Prestige and Willshaw, 1975; Schmidt and Easter, 1978).
3. Axon *branching* (in the form of interstitial branching, backbranching, growth cone bifurcation, or terminal arborization) contributes to retinotectal map formation, although to different degrees in different model organisms (Fujisawa *et al.*, 1982; Kaethner and Stuermer, 1992; McLaughlin and O'Leary, 2005; O'Rourke and Fraser, 1990; Roskies and O'Leary, 1994; Simon and O'Leary, 1992).

Spontaneous or correlated waves of neural *activity* in the retina and tectum contribute to map formation as well, though this is more of a secondary role. Generally speaking, activity is considered to carry out a refinement of the projection once the preceding mechanisms (above) have generated an initial (coarse) map (Cang *et al.*, 2008b; Debski and Cline, 2002; McLaughlin *et al.*, 2003; O'Rourke *et al.*, 1994; Ruthazer and Cline, 2004; Ruthazer *et al.*, 2003; Schmidt, 1990).

The four mechanisms already listed are well established in retinotectal map formation and are supported by a wealth of evidence. An additional hypothesized mechanism is *marker induction*, which refers to the respecification of tectal (and possibly retinal) cues during map development and regeneration. It is most commonly used to mean respecification of tectal markers due to the presence of RGC axons, and is hence often referred to as retinal induction. It was first formalized as a hypothesis for modeling in topographic map formation by von der Malsburg and Willshaw (1977). Early data suggested such a mechanism might be at work (Gaze *et al.*, 1974; Schmidt, 1978; Sharma, 1972; Yoon, 1976, 1980) but molecular demonstrations of this respecification have been limited (King *et al.*, 2003; Rodger *et al.*, 2000). There are other mechanisms that could potentially impart topographic information, such as the order of fibers within the optic nerve/tract and the temporal order of arrival of fibers on the tectum, but they do not in fact appear to make an important contribution (Holt, 1984; Stuermer, 1986).

3.2. Systems-based modeling work

Following the proposition of Sperry (1963), experiments designed to test the contribution of chemoaffinity to retinotopy were performed and were accompanied by the first formal quantitative models of map formation in this context. This generation of "systems level" experiments involved various surgical manipulations of both retina and tectum to test the chemoaffinity hypothesis and how robust the retinotectal system as a whole was to these perturbations. An illustrative example of the many experiments done in this vein are the retinal ablation (or map expansion) experiments. In one such example of these experiments (Schmidt *et al.*, 1978), half-retinal ablations were performed on goldfish eyes and the resulting map observed. In some cases, the remaining retinal fibers remained confined to their appropriate hemitectum, but it was found that they eventually shifted their connections so that the remaining fibers expanded their projection to fill the entire tectum (Fig. 1.6). These and other experiments (Goodhill and Xu, 2005; Udin and Fawcett, 1988) clearly demonstrated that while chemoaffinity is important in retinotectal mapping, there was significant plasticity in the map that could not be accounted for by chemoaffinity alone. Hence successful modeling relied on incorporating multiple mechanisms or constraints to recreate the experimental results.

3.2.1. Competition and types of affinity

One of the first computational studies on the formation of retinotectal map (Prestige and Willshaw, 1975) demonstrated that assumptions underlying chemoaffinity needed to be clarified, and introduced a classification of type I and type II chemoaffinity mechanisms. The authors considered a mapping between a pair of one-dimensional arrays of cells; each projecting cell was allowed to make multiple contacts (a representation of branches) in the target cell array, and target cells were similarly each allowed to accept multiple contacts. Strengths of connections were expressed as lifetimes of the contact (cf. synaptic strength). Type I chemoaffinity (or "rigid matching") assumes that each projecting cell has maximum affinity for only one (or other small subset) of the target cell array, and each cell in the target similarly has maximum affinity for only a subset of projecting cells. Type II matching describes *graded* affinity, whereby all projecting cells have maximum affinity for one end of the target (e.g., rostral tectum), and all cells in the target have maximum affinity for one end of the projecting cell array. Type I mechanisms can certainly form an ordered map, but not as immediately clear was whether type II mechanisms could do the same, and under what conditions. In simulations the authors were able to show that type II mechanisms were indeed able to form a retinotopic map, but only if limits were placed on the number of contacts a projecting cell could form with target cells, and vice versa. It was observed that this constraint could be realized in, for example, competition for a target resource such as space ("competition by exclusion"). The authors acknowledged that while this model could form normal maps and certain systems manipulations, others (such as map expansion and compression) required further assumptions that were not well justified. Although the simulations were only carried out in one dimension, the ideas easily extend to two; as such, competition has been a common feature of retinotectal models from this time on.

3.2.2. Marker induction models

Type I matching has the advantage of a more straightforward implementation and does not require competition to form an ordered map, but is problematic in that it cannot alone account for the plasticity of maps demonstrated in systems manipulations. To account for these results in a type I scheme, another mechanism such as marker induction needs to be employed. It was this idea was explored computationally in a series of models (von der Malsburg and Willshaw, 1977; Willshaw, 2006; Willshaw and von der Malsburg, 1979) best described as marker induction models (aka Tea Trade model, retinal induction model). These models connected retinal axons to tectal cells by contacts or synapses, whose weights were modulated depending on the similarity of markers between the projecting cell and the receiving cell. New contacts could be formed near other contacts in areas more likely to yield successful connections, and competition was introduced by normalizing the synaptic output of RGCs. Induction occurs as markers either diffuse out of (von der Malsburg and Willshaw, 1977; Willshaw and von der Malsburg, 1979) or are upregulated by (Willshaw, 2006) RGC axon terminals; neighboring tectal cells can also influence each other through this effect. Coupled differential equations were defined for changes in marker level and synaptic strength. Through this mechanism, markers on small regions of tectum become similar to markers on small regions of retina, which can lead to topography. Willshaw (2006) updated the model to include newer molecular data (see Section 3.3), and also measured map precision in terms of mean receptive field size and separation.

The marker induction model can be described as a type II chemoaffinity model, which also employs competition and branched axons. This was the first use of branching axons (albeit abstract ones) as a kind of pathfinding device in retinotectal modeling. The synaptic modulation approach used is similar to Hebbian models of activity-dependent synaptic plasticity, but uses similarity of markers instead of correlations in activity. (Indeed an activity-based version of the model subsequently formed the basis for the elastic net model (Durbin and Willshaw, 1987), which had comparatively greater success in influencing modeling cortical feature maps; this is discussed further in Section 4.) An important property of this model that was demonstrated in simulations (and has been found in subsequent models employing similar activity-based Hebbian rules) is that it cannot generate topography without some sort of overall polarity guide (such as a weak initial gradient of markers). Although some of the assumptions of the model, particularly details of the induction mechanism, require more experimental justification, these same assumptions give it a great deal of flexibility. It is able to explain a wide range of systems-based results (see, e.g., Fig. 1.7), and is also one of the few models that has also been applied to more recent molecular data (Willshaw, 2006).

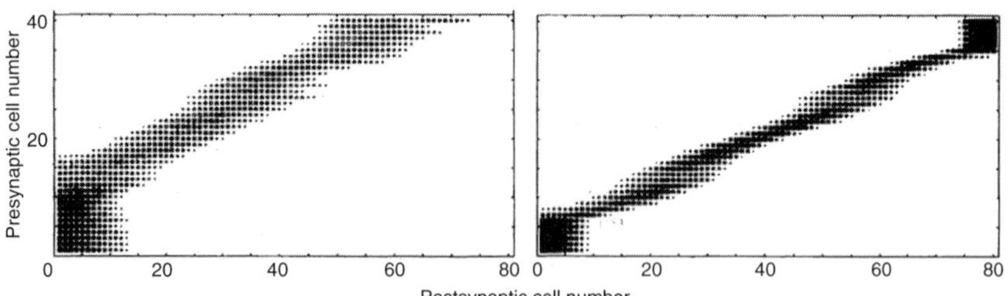

Figure 1.7 Simulation of retinal ablation experiments using the marker induction model. RGCs retinal position plotted against tectal position. Each dot represents a synapse between the corresponding presynaptic and postsynaptic cell, with area proportional to synaptic strength. The left panel shows the simulation early after retinal ablation, where the RGCs retain some of their original ordering prior to ablation. The panel on the right shows the final positions, replicating the map expansion results seen in experiments (from Willshaw and von der Malsburg, 1979; reproduced with permission from the Royal Society).

3.2.3. Sorting-based models

The Arrow model (Hope *et al.*, 1976) took a more abstract approach to modeling axon–axon interaction, treating it as a kind of sorting mechanism. An initially unordered array of RGC axon terminations on a tectal grid was considered. Pairs of neighboring array members were compared and their positions exchanged if they were incorrect with respect to both their retinal origin and orientation compared to local tectal polarity or "arrow." This exchange process (a sort of type II affinity) could be iteratively alternated with a random walk step to vacant array sites, although this was not done in the simulations presented. The model was able to generate ordered maps and recreate some systems data, such as tectal graft rotation experiments, but not tectal graft translocation experiments.

To account for the remainder of the systems data Overton and Arbib (1982b) updated the Arrow model to XBAM (eXtended Branch Arrow Model), which in itself was an updated version of a previous model (BAM) by the same authors (Overton and Arbib, 1982a). There were a number of significant changes; firstly, the tectum was treated as continuous, and the Arrow model exchange rules converted to movement vectors induced in RGC axons by other nearby axons. The boundaries of the tectum as well as graft edges also contributed to the movement vector. A softened type I affinity rule was added to reflect results of translocation experiments, so that the model now employed a mixed type I/II affinity. Each axon consisted of a number of branches, and it was assumed that movements of branches from the same axon were allowed to influence each other through an averaging process (independent of their proximity to each other). XBAM successfully explained a wide range of plasticity results through its multiple constraint approach and mixed type I/II affinity. Its consideration of branched axons and motion over a continuous tectum were also quite advanced for its time in terms of realism.

3.2.4. Dual gradient chemotaxis models

A more explicit use of molecular marker ideas, and a prototypical type I affinity model, was proposed by Alfred Gierer in a series of models in the 1980s (Gierer, 1981, 1983, 1987). These models considered molecular gradients (which were largely speculative at the time) in the retina and tectum, which exerted their effects by producing a substance p within individual growth cones, which is in general a function of retinal origin and current tectal position. It was proposed that growing axons seek a maximum (or minimum) of p, by moving in the direction where the rate of change of p was maximal (or most negative). To achieve the required maxima and minima, dual gradients (generally in opposite directions) were required. These are sometimes known as "countergradients," though this term has also been used to describe pairs of gradients oriented in opposite directions, where one gradient is in the retina and the other is in the tectum. Acknowledging the inability of this type I mechanism to reproduce plasticity phenomena in the map, Gierer (1983) included a form of retinal induction ("regulation") and used this to explain some systems experiments (compression and expansion) but not others (e.g., translocations, rotations, etc.). Later versions of the model (Gierer, 1987) presented limited simulations of how branching and axonal trajectories might be expected to follow from *p*-guidance model (Fig. 1.8), but did not include regulation.

These models were unusually realistic in their consideration of responses to gradients (they are similar to more modern axon guidance models), branching, and trajectories. However, these concepts were not combined into a cohesive model in simulations of map development, so that only limited experimental data was replicated. Notably though, realistic branching patterns and trajectories were reproduced, which has been unusual in the modeling literature. The use of dual gradients represents a rigid matching approach, which is unable to account for many systems experiments,

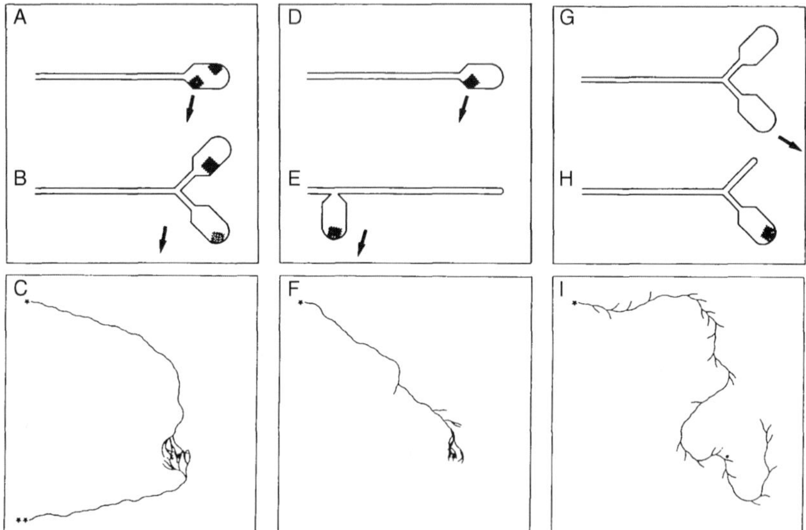

Figure 1.8 Gierer's branching and trajectory simulations. Top row: Examples of how branching might occur using the p-guidance model. Bottom row: simulations using these models. (A–C) If the growth cone encounters a negative slope (below a certain threshold), it becomes "proximally activated" and initiates "forward branching" (or growth cone bifurcation). In (C), two axons are simulated to show the different trajectories that can be observed from different starting points (filled stars) but the same target (filled circle). (D–F) Similar to the above, but proximal activation of the growth cone now causes axon growth to stop, and branching to occur behind the growth cone (backbranching). (G–I) Now branching occurs randomly, and lateral inhibition between branches causes the branch with the higher level of p to win, while the other branch stops growing (from Gierer, 1987; reproduced with permission from the Company of Biologists Ltd.).

despite the inclusion of a form of marker induction to remedy this. There are also constraints on the gradient shapes that can form maps under the proposed scheme; that is, ones for which $dp/dx = 0$ implies retinotopy (Goodhill, 1998). Despite these drawbacks, Gierer's models provide a relatively detailed and realistic framework, and their general assumptions form the primary example of what are known as *dual gradient* or *countergradient* models.

3.2.5. Adhesive energy minimization models

The idea of minimizing a certain quantity to form a map formed the basis of a series of models based around the adhesive properties of retinal and tectal cells (Fraser, 1980; Fraser and Perkel, 1990). Disks representing terminal arbors were arranged in an initially random configuration on the tectum. An "adhesive free energy" was calculated for each disk in its current location based on contributions from molecular markers, correlated activity, and competition (hence it is also known as Fraser's multiple constraint model). The disk was randomly moved to a new position, and if this resulted in a reduction in free energy, the move was accepted; otherwise further random movements were attempted. Through this approach, the map was evolved to stability.

Similarly to Gierer's model, gradients of molecular markers in a type I affinity scheme were used. In addition, two separate axon–axon interactions were modeled; one activity-based, and the other a chemospecific interaction whereby axons with similar levels of markers had a lower adhesive energy associated with the interaction (cf. retinal induction models). This latter property, in the context of this model, represents a type II mechanism. Hence, the model is a multiple constraint model with mixed type I/II affinity. It was able to successfully simulate a wide range of experimental results (Fig. 1.9), but the explanatory power of the model was reduced by the abstract nature of rules for movement and minimization.

3.2.6. Chemospecific synaptic modulation

A series of models developed by Cowan and colleagues were largely based on synaptic modulation rules, in which the change in strength of a synapse between a pair of retinal and tectal cells is considered to be proportional to some combination of chemospecific adhesiveness between them and the correlation in their activity (Cowan and Friedman, 1990, 1991; Weber *et al.*, 1997; Whitelaw and Cowan, 1981). Although models in this series had much in common, each focused on different elements in their implementations. Whitelaw and Cowan (1981) modeled the growth of the synapse as proportional to the product of chemospecific adhesion and correlated activity and used a type II chemoaffinity. Cowan and Friedman (1990, 1991) added axon–axon interactions and changed from type II to type I affinity. Weber *et al.* (1997) made the chemospecific and activity-based contributions independent, and added two axon–axon interactions in a multiple constraint

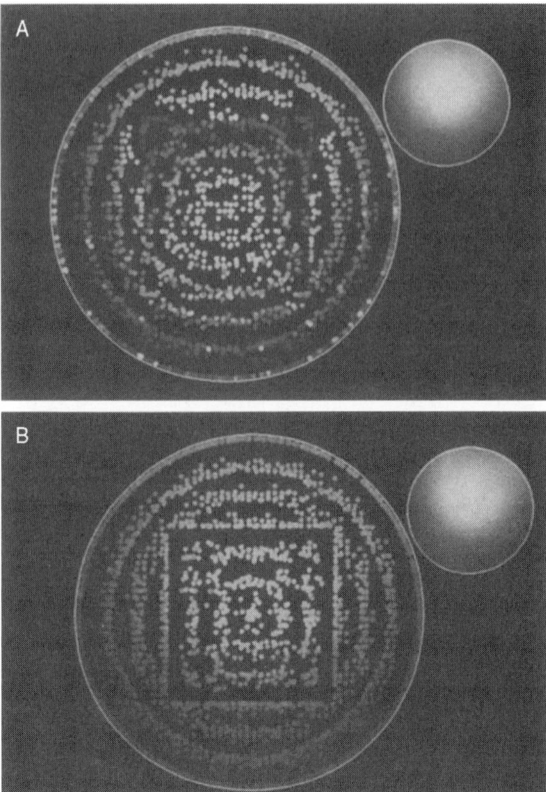

Figure 1.9 Simulation of tectal rotation experiments in the adhesive energy minimization model. The two smaller circles correspond to the retina, while the larger circles represent the tectum. Locations of RGC arbors are indicated by dots within the larger circle. In (A) a rotated map forms, following the rotated tectal cues, while in (B) a normal map results that ignores the rotated cues. The normal map tended to form when the degree of initial order was high (from Fraser and Perkel, 1990; reproduced with permission from John Wiley & Sons, Inc.).

framework, similar to the energy minimization models discussed above. These models, and in particular Weber *et al.* (1997), were quite detailed and successful in explaining experimental results, but were limited in realism by not having specific mechanisms for growth and movement of axons.

3.3. Chemoaffinity and the ephrins

A new wave of enthusiasm in experimental and theoretical exploration of the retinotectal map began when molecular effectors for Sperry's hypothesized gradients were discovered and characterized in the mid-1990s.

The primary example of these molecules are the Eph receptors and their ligands the ephrins (although other molecules may also contribute to some extent). Gradients of Eph receptors in the retina and ephrin ligands in the tectum are matched as suggested by Sperry, and were shown to be intimately involved in retinotopy. (For reviews of the contributions of the Eph–ephrin molecules to retinotopic mapping and many other areas of biology, see Flanagan and Vanderhaeghen (1998), Halloran and Wolman (2006), Kullander and Klein (2002), McLaughlin and O'Leary (2005), Pasquale (2005), Poliakov *et al.* (2004), and Wilkinson (2001).)

The discovery of the Eph and ephrin gradients led to a spate of sophisticated experiments to characterize their contribution to retinotectal mapping (a simple example is shown in Fig. 1.6), while new models were created and old ones modified to grapple with the data. We suggest that, ironically, this resulted in a step back in theoretical understanding, as all mechanisms except chemoaffinity were de-emphasized, and sometimes ignored completely. As a result, models of this generation generally have not replicated the systems-based results of the previous generation. Indeed certain features of the newer molecular-based experiments remain unexplained by these recent models as well, so that incorporating features from earlier models may be of benefit (e.g., see Goodhill, 2000; Wilkinson, 2000).

3.3.1. Servomechanism models

The servomechanism model (Honda, 1998, 2003, 2004) is a computational model based on the observations by Nakamoto *et al.* (1996). The model assumes that a signal S is generated in growth cones, which is a function of its fixed receptor level R (determined by retinal soma position) and local ligand level L (determined by current position on the tectum). This was expressed as $S = R \cdot L$, using mass action principles at equilibrium, and it was proposed that RGC axons seek out a standard signal S_0, by comparing the current signal S with the standard value. Starting at the rostral tectal border, axons were chosen at random to be moved in a probabilistic fashion to a new position. The chance of moving to a new position is higher if that position reduces the value $|S_0 - R \cdot L|$, and greater reductions in this value give greater probabilities of moving to the new location. This, combined with a general tendency for axons to move caudally and dorsally, allowed for topographic map development. The "set point" rule employed here is a type I affinity model in the mould of Gierer's models (Section 3.2.4), and is not easily able to recreate systems experiments. As expected, another mechanism was required to demonstrate plasticity in the mapping. Honda (2003) added competition in the form of a rule for moving axons away from more dense areas to less dense areas, until the density of terminations was uniform or below a critical density n_c (although this value was set to zero in Honda (2004) so that competition takes place everywhere). The axons that are chosen to be moved away from more dense areas to less dense areas are those that have with the smallest $|S_0 - R \cdot L|$, so that this competition step also involves some sorting. This competition was computationally separate from the servomechanism phase; that is, first an initial mapping is made with the servomechanism rules, then the density of terminals is evened out with the competition rules. The model was then able to recreate some limited systems manipulations (Honda, 1998, 2004) and some ephrin misexpression data (Honda, 2003). Variations of the model have been considered by Löschinger *et al.* (2000) and Thivierge and Balaban (2007).

The servomechanism model is the only modern model to attempt to simulate some limited systems-based data in addition to the newer molecular data, and the only computational model that has been applied to stripe assay results (see, e.g., Fig. 1.10). Effectively, the servomechanism model involves a noisy gradient ascent algorithm (including a stop signal), followed by a smoothing/sorting algorithm (i.e., type I followed by type II affinity). This scheme, where an initial chemoaffinity phase is followed by a spreading

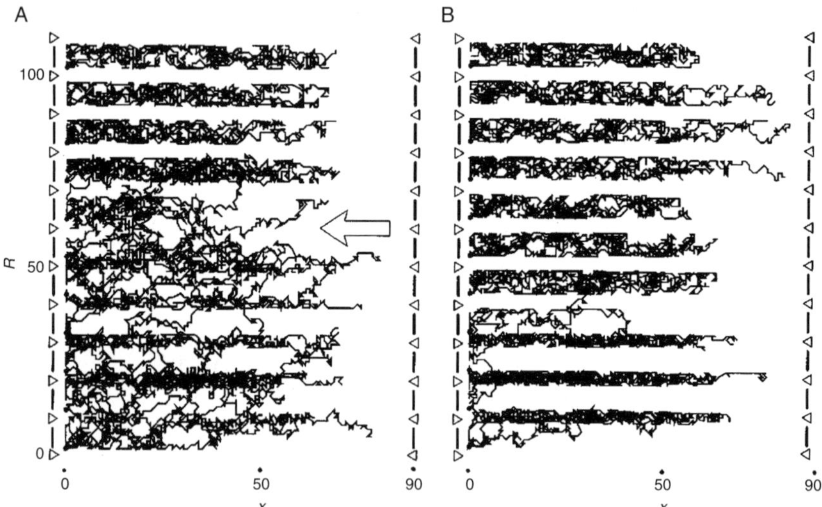

Figure 1.10 Servomechanism model simulations of a stripe assay. In this particular kind of stripe assay, alternating stripes of anterior and posterior tectal membranes are laid down, and axons from nasal to temporal retina are placed at one end (Walter *et al.*, 1987a, b). Here, anterior stripes are indicated by short vertical bars and posterior stripes are marked by open triangles. The vertical axis shows receptor level R, which corresponds to position of retinal origin, and the horizontal axis represents the distance x that axons have "grown" along the stripe. (A) Stripes composed of "crude" membrane fragments are simulated (i.e., differences in ligand level between stripes is small). (B) Stripes composed of fractionated membrane fragments (large difference in ligand between the two stripes). Experimentally a mix of striped and nonstriped growth is seen; usually from temporal and nasal axons, respectively. Sometimes sharp transitions from striped to nonstriped growth are seen, and a possible correlate of this in Honda's simulations is shown by the large open arrow (from Honda, 1998; reproduced with permission from Elsevier).

out of connections, was initially postulated by Prestige and Willshaw (1975). Although the use of type I affinity followed by competition gives the model some flexibility, this type of sequential process is somewhat artificial, and is not biologically well supported. There is some evidence for local maxima of adhesion, cell attachment, and/or outgrowth in gradients (Hansen *et al.*, 2004; Huynh-Do *et al.*, 1999), but it is not clear whether or how this translates into a "set point" rule (or stop signal) *per se*; at least not without further assumptions. Some of the movement rules, such as the tendency for dorsocaudal movement, also represent rather arbitrary assumptions. These issues aside, the model produces some detailed simulations which are comparable with a wide variety of experimental data.

3.3.2. Probabilistic sorting models

Koulakov and Tsigankov developed models involving probabilistic sorting of positions in a discrete array, similar to the sorting approach of the Arrow model, but now with explicit reference to Eph–ephrin gradients (Koulakov and Tsigankov, 2004), and later also including activity (Tsigankov and Koulakov, 2006). From initially unordered conditions, a randomly chosen pair of axons is considered and their positions exchanged if their current ordering is incorrect with respect to the direction of tectal gradient and relative retinal position. Although neighboring axons were considered in one-dimensional versions of the model, two-dimensional versions exchanged axons that were not necessarily neighboring. Differing probabilities are assigned to this exchange based on absolute differences in Eph/ephrin level, and whether the interaction is attractive or repulsive. Tsigankov and Koulakov (2006) included activity as well as this type II affinity, and a different algorithm was used for calculating exchanges. Dual gradients (countergradients) were included implicitly through functional inactivation or "masking" (Hornberger *et al.*, 1999). Repulsive and attractive gradients of Eph and ephrin molecules (in the rostrocaudal and mediolateral tectal axes, respectively), and correlated activity were assumed to contribute to a form of adhesive energy. Pairs of axons were again considered, and their positions exchanged if this reduced the adhesive energy. This equates to minimizing repulsion, while maximizing attraction and correlated activity, and hence bears resemblance to the adhesive energy minimization approach in Section 3.2.5. Calculations were presented to analytically predict features of the doubled and collapsed maps of Reber *et al.* (2004), making this one of the few models to present analytical arguments in addition to simulation data.

The use of type II affinity with implicit competition (and in the later version, activity) gives this two-dimensional model enough constraints to be applied to various experimental results, but so far only some ephrin misexpression results have been explored (Fig. 1.11). The authors used arguments based on noise in gradient-sensing and the spatial limits of activity-based mechanisms to explain different experimentally observed phenomena.

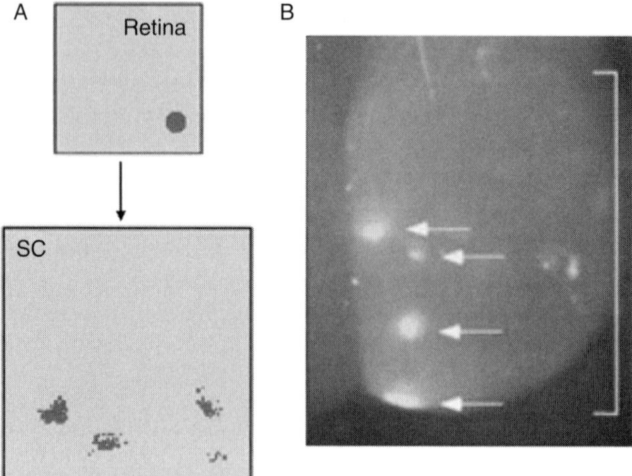

Figure 1.11 Simulations and experimental results in ephrin misexpression studies. (A) Simulated injection of tracer dye projects to the superior colliculus (SC) in the probabilistic sorting model (Tsigankov and Koulakov, 2006) applied to an ephrin misexpression study. (B) Corresponding experimental results for the ephrin misexpression study (Feldheim *et al.*, 2000) where focal retinal injections of DiI into the retina (similarly to the simulations shown in (A) but the experimental retinal injection is not shown) resulted in multiple termination zones (arrows) in knockout animals ((A) from Tsigankov and Koulakov, 2006; adapted with permission from Springer Science + Business Media. (B) from Feldheim *et al.*, 2000; reproduced with permission from Elsevier). (See Color Insert.)

Although they are commonly used mathematical tools, minimization procedures in the context of retinotectal mapping have the problem that there does not appear to be a simple biological explanation for how this minimization is performed; in particular, how neural activity and molecular cues can both contribute to an energy that can be minimized by the growth cone. It is also not clear how actual movement results from this kind of scheme, nor how non-neighboring axons can interact and exchange positions in all circumstances (although the effect of this latter point was considered in Koulakov and Tsigankov (2004)). Despite these limitations in terms of biological realism, this probabilistic sorting model represents a concerted effort to engage with the ephrin misexpression literature, both in simulations and using more analytical methods.

3.3.3. Dual gradient branching model

A rather different approach was taken in the computational model of Yates *et al.* (2004) in which the focus was placed on biased branching of RGC axons in generating topography. This model followed the observations that branching along the length of an axon that initially overshoots its target appeared to be particularly important in mammalian and avian topographic map development (Roskies and O'Leary, 1994; Yates *et al.*, 2001). The particular pattern of interstitial branching observed is regulated in part by Eph–ephrin interactions (McLaughlin and O'Leary, 2005; Rashid *et al.*, 2005; Simon and O'Leary, 1992; Yates *et al.*, 2001). In this model, dual gradients of Eph–ephrins were assumed to be present in both retina and tectum (a type I affinity model, with assumptions similar to Gierer's). Probabilistic rules were introduced which allowed branching to occur below a certain threshold of repulsion, and set branching to be maximal where repulsion was minimal. Additional rules allowed advancing, pausing, and retraction of branches. New branches contributed additional molecular markers to the tectal gradients (hence axon–axon interactions are incorporated implicitly), in some cases sharpening the gradients and therefore the resulting map. To obtain a better match of the behavior of the model to biology, an extra "branch density" term needed to be added to the model, which was time-varying and acted to increase branching probability in areas of already high branch density.

Simulations of ephrin misexpression studies were presented and are generally comparable to the results seen in experiment. However, this model does not attempt to address the inability of its type I assumptions to explain plasticity results (unlike Gierer's model which included regulation, and the servomechanism model which includes competition). One of the basic assumptions of the model, that branching is inhibited by repulsive markers on the tectum and from added branches, seems to be in contradiction with the assumption of the added "branch density" term, that branching is *promoted* by the presence of high densities of branches (i.e., high density of repulsive markers). It is also unclear how this time-varying extra term relates to the biology (e.g., activity-based mechanisms as suggested by the authors), how the specific choices of free parameters (e.g., threshold for branching) affect modeling outcomes, or whether the use of other types of gradients as suggested (e.g., dual attractive or mixed attractive and repulsive) could operate in the same way. However, this model was the first to show that branching alone, biased by gradients of molecular markers, is in principle capable of generating retinotopy.

3.3.4. Relative signaling models

Reber *et al.* (2004) presented a model based on quantitative descriptions of EphA misexpression data (Brown *et al.*, 2000; Reber *et al.*, 2004) showing that relative, rather than absolute, levels of EphA were important in mapping. This relative signaling was proposed to occur through local comparisons between RGC axons (a type II affinity), and it was assumed that a discrimination limit of relative signaling operated in this context. Calculations based on these concepts accurately predicted the position of map collapse (or lack thereof) in different phenotypes (Fig. 1.12).

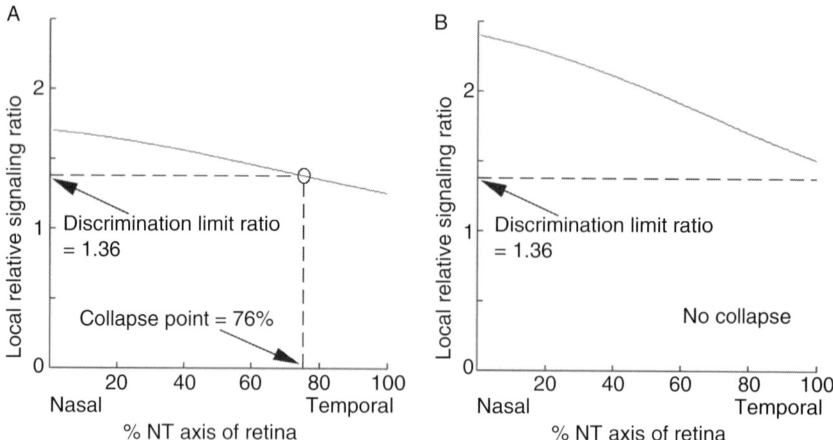

Figure 1.12 Relative signaling discrimination limit. In Brown *et al.* (2000) and Reber *et al.* (2004), the level of Eph receptor was artificially increased in half of all RGCs (distributed randomly throughout the retina). In some cases, the knockin RGCs display a second "doubled map" in addition to that formed by the wild-type RGCs. Here, the relative signaling ratio (the ratio of Eph receptor level on wild-type RGCs to receptor level of knockin RGCs) is plotted against retinal position. (A) The heterozygote case where the doubled map collapses to a single map at 76% of the map axis (giving a predicted relative signaling ration of 1.36). (B) The homozygous case shows no collapse of the doubled map using the same ratio, and this mirrors the experimental outcome in homozygotes (from Reber *et al.*, 2004; adapted with permission from Macmillan Publishers Ltd.).

To account for wild-type maps where the relative signaling ratio for neighboring RGCs is ≈ 1, and well below the calculated discrimination limit of 1.36, a mechanism of comparison with a universal reference level was invoked. Although no actual simulations of map development were performed, suggestions were presented for how relative signaling may be carried out in this context.

The relative signaling model replicates specific important results that other models have been largely unable to account for, particularly the presence or absence, and position of, map collapse points. Certain features of how this algorithm leads to mapping are problematic though, such as the idea of a comparison with a universal reference in wild-type maps, as this does not appear have a simple biological explanation. It is possible that including additional constraints, such as competition, could avoid the need for this assumption.

3.4. Summary of models

Table 1.1 summarizes the main types of mechanism used by the models discussed above.

Table 1.1 Model features summary table

Model	Version	Chemoaffinity	Branching	Competition	Activity	Induction
Competitive affinity	1975	Type I/II	–	Implicit	–	–
Marker induction	All	Type II	Exploratory	Explicit	–	Yes
Sorting-based: Arrow	1976	Type II	–	Implicit	–	–
Sorting-based: BAM	1982a	Type II	Averaging	Explicit	–	–
Sorting-based: XBAM	1982b	Mixed I/II	Averaging	Explicit	–	–
Dual gradient chemotaxis	1983	Type I	–	–	–	Yes
	1987	Type I	Exploratory	–	–	–
Adhesive energy minimization	All	Mixed I/II	–	Explicit	Yes	–
Chemospecific synaptic modulation	1981–1991	Type II	–	Explicit	Yes	–
	1997	Type I	–	Explicit	Yes	–
Servomechanism	1998	Type I	–	–	–	–
	2003, 2004	Mixed I/II	–	Explicit	–	–
Probabilistic sorting	2004	Type II	–	Implicit	–	–
	2006	Type II	–	Implicit	Yes	–
Dual gradient branching	2004	Type I	Exploratory	–	–	–
Relative signaling	2004	Type II	–	–	–	–

The main mechanisms used in models of retinotectal mapping, and the computational models that use them. Models are referred to by their main mechanism as described in the text. Different versions of the same model series are referred to by year; each corresponds to the citations in the main text. Implicit competition refers to models where competition is implied by the assumptions of the model (e.g., in 1–1 mappings) rather than described by explicit equations.

4. Activity-Dependent Development

Theoretical models for the development of activity-dependent wiring generally assume that an initial connectional architecture has been established by molecular cues, using the mechanisms discussed in the previous sections. The models then posit "learning rules" for how synaptic strengths change within this fixed architecture. Usually, these rules specify how particular combinations of presynaptic and postsynaptic activity lead to incremental increases or decreases in the synaptic strength, or "weight," of the connection. It is generally assumed that these changes lead to the organism becoming better adapted to its environment than if all synaptic strengths were fixed in advance. Historically this formalism has been shared with the field of artificial neural networks, which experienced a strong resurgence of interest in the 1980s (e.g., Hinton, 1989, 1992). Sometimes the goal of developing ANNs which can solve difficult practical learning problems in the fields of, for instance, perception, classification, and action selection has been taken as synonymous with understanding how nervous system wiring develops. While there has certainly been a rich flow of insight in both directions, here we will mostly focus on theoretical models which engage more directly with neuroscientific data (Dayan and Abbott, 2001).

4.1. Types of learning

From a computational point of view, it is common to distinguish three types of learning: supervised, reinforcement, and unsupervised learning. In *supervised* learning, the goal is to learn a mapping between given input and output vectors, for instance classifying the identity of faces in a set of images. This requires a way of propagating information about errors in the output for a given input through the network so that the strength of all relevant synapses can be changed appropriately. While over the past several decades powerful methods have been developed for such learning (e.g., Ackley *et al.*, 1985; Hinton, 2007; Minsky and Papert, 1969; Rumelhart and Zipser, 1986), the relevance of these to experimental data regarding neuronal wiring development is only at an abstract level. It is also hard to imagine how the detailed "teaching" signal that such algorithms require to provide error information could be present biologically. However, it remains possible that some of the computational principles underlying effective learning identified by these approaches might in the future find direct application for understanding biological wiring development.

In *reinforcement* learning, the goal is to learn a mapping between a set of inputs or actions in a particular environment and a (possibly temporally delayed) scalar-valued output representing some measure of "reward." Computationally this is much harder than supervised learning, since it requires a way of assigning credit or blame to individual actions which may have occurred some time before the reward signal arrives. Good progress has been made on this problem computationally through the development of methods such as temporal difference learning (Sutton, 1988), and there is direct evidence that such algorithms are indeed employed in reward processing in the adult nervous system (e.g., Schultz *et al.*, 1997). The extent to which these algorithms can be used to understand nervous system development remains an intriguing unanswered question.

In *unsupervised* learning, the network is provided with no feedback at all. Rather, synaptic strength changes occur according to a learning rule based only on pre- and postsynaptic activity, with no explicit specification of the desired or "best" output for each input. The outcome (i.e., the pattern of synaptic strengths developed) depends on the nature of the learning rule and the statistical structure of the inputs presented. For example, the learning rule might imply finding clusters in the inputs, without specifying what meaning the clusters found should have for the animal. This approach can be very effective for understanding the development of neuronal wiring patterns at early stages of sensory processing, where it is reasonable to hypothesize that representations are driven more by just the statistical structure of the input data than the specific goals of the animal. Indeed, unsupervised learning methods have proven highly successful at reproducing the development of early stages of visual processing, as we discuss in detail below.

4.2. Linear Hebbian learning

A particularly fruitful hypothesis for how synaptic strengths change during both development and learning is the rule of Hebb (1949):

> When an axon of cell A is near enough to excite a cell B and repeatedly or persistently takes part in firing it, some growth process or metabolic change takes place in one or both cells such that A's efficiency, as one of the cells firing B, is increased.

A direct biological implementation of such "coincidence detection" emerged with the discovery of long-term potentiation (LTP), first in the hippocampus and subsequently in other brain regions including the cortex (Malenka and Bear, 2004). Qualitatively, Hebbian-type rules appear to explain many phenomena in neural wiring development; for instance the development and plasticity of ocular dominance columns in primary visual cortex (Katz and Shatz, 1996). Quantitatively, there are many different mathematical equations consistent with Hebb's qualitative statement. A large amount of research in models of neural wiring development has been devoted to understanding the outcomes of different types of Hebbian equations, and how to relate these outcomes to biological phenomena.

Although historically the first models were nonlinear, it is easier to explain some basic concepts in the context of linear models. We will first present in some detail a simple but highly illustrative example: a two-layer network with a single output neuron and a linear activation rule (e.g., see Dayan and Abbott, 2001; Hertz *et al.*, 1991).

In particular consider the simple network shown in Fig. 1.13. We assume the output activity y is given by

$$y = \sum_i w_i x_i, \tag{1.1}$$

where the w_is are the weights and the x_is are the inputs. This can be expressed more compactly in vector form as $y = \boldsymbol{w} \cdot \boldsymbol{x}$. In this model, the detailed temporal dynamics of neural activity are not considered; instead, the focus is on the timescale of weight changes, which is here assumed to be much slower. One of the simplest mathematical expressions of Hebb's rule is

$$\Delta \boldsymbol{w} \propto y \boldsymbol{x}, \tag{1.2}$$

which says that a small change in the weights occurs which is proportional to the product of the output activity and the input activity. A slightly more subtle and biologically realistic version (Sejnowski, 1977) is

$$\Delta \boldsymbol{w} \propto (y - \langle y \rangle)(\boldsymbol{x} - \langle \boldsymbol{x} \rangle),$$

where the angle brackets denote time averaging. In this equation, we have subtracted the average value from both the input and output activities. We can now combine this with Eq. (1.1), take a time average over all possible input patterns, and assume that the weights change slowly relative

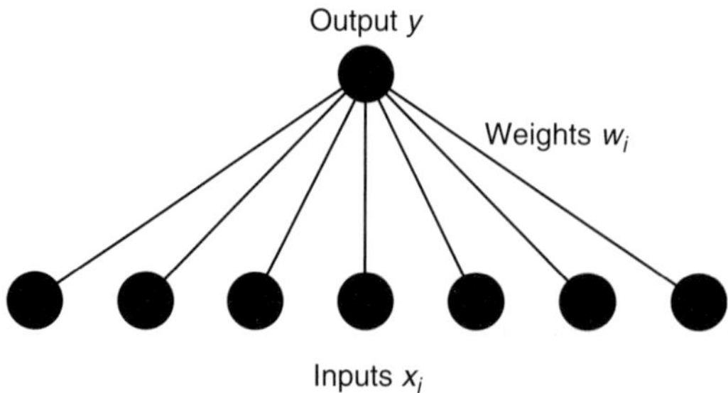

Figure 1.13 A simple neural network. Activity x_i in input neurons i is conveyed by fibers with synaptic weights w_i to an output neuron with activity y. A linear Hebbian learning rule for the weights w_i is sufficient to produce interesting behavior of the network.

to the rate at which input patterns are presented. That is, what matters for the weight dynamics is not the properties of any one individual input pattern, but rather the average statistical properties of the entire set of input patterns. This leads straightforwardly to the following equation:

$$\langle \dot{\boldsymbol{w}} \rangle = \alpha \mathbf{Q} \boldsymbol{w}, \qquad (1.3)$$

where $\mathbf{Q}$ is the covariance matrix of the inputs.

This differential equation specifies temporal dynamics for the weights: how they change in response to the average correlations in the inputs. It is mathematically straightforward to immediately predict the outcome of this dynamics by imagining the weight development in the basis of the eigenvectors of $\mathbf{Q}$. Since the eigenvectors are the set of axes along which each component grows independently, with rate determined by the eigenvalue for each axis, eventually the eigenvector with the largest eigenvalue will dominate. That is, $\boldsymbol{w}$ tends to the principal eigenvector of $\mathbf{Q}$. Remarkably, this simple rule achieves a sophisticated computational goal. The principal eigenvector (also known as the principal component) of $\mathbf{Q}$ represents the direction along which the input data is most spread out in the sense of having maximum variance (e.g., Krzanowski, 1988). That is, the Hebbian rule above naturally tends to find patterns of weights which produce the biggest range in activity for the output neuron, given the statistical structure of the inputs presented. Even more remarkably, this learning rule also turns out to maximize the amount of information about the input that is conserved in the output (Linsker, 1990).

4.3. Constraining the weights

A problem with the simple Hebbian rule described above is that it is unstable: eventually all the weights go to infinity. One way to avoid this is to add extra terms to the right-hand side of Eq. (1.3). An example is Oja's rule (Oja, 1982, 1989), which again finds the principal component of the data but maintains the weights within bounds. A second method is to simply saturate the weights at upper and lower limits, which can be thought of as constraining the weights inside a box in the weight space. However, this makes it much harder to analyze the learning rule, since once the weight vector hits the walls of the box its behavior will be highly nonlinear (Feng *et al.*, 1997). A third method is weight normalization, which is motivated by the idea that the total strength of synaptic inputs that a postsynaptic neuron can support depends on some limited resource (reviewed in Miller, 1996). The most common way to implement this is to assume that either the sum of weights or the sum of squares of weights is maintained at a limiting value. The latter is equivalent to saying that the weight vector is constrained to be of constant length. Whichever normalization constraint is used, there are two common ways to enforce it (Goodhill and Barrow, 1994; Miller and MacKay, 1994). In divisive (aka multiplicative) enforcement each weight is scaled proportional to its size, while in subtractive (aka additive) enforcement the same amount is subtracted from each weight to enforce the constraint.

4.4. Application of linear Hebbian learning to visual system development

Hebbian learning rules very closely related to those described above have been used to model both ocular dominance and orientation column development in the primary visual cortex. Linsker (1986) used a generalized form for Eq. (1.3) in a model consisting of a series of layers, with learning in each layer driven by correlations in the layer below. Neurons were assumed to have spatially localized receptive fields, and random activity (white noise) was applied to the input layer. This simple model was capable of producing several different types of receptive fields including all-excitatory, all-inhibitory, center-surround, and oriented. Furthermore, adding fixed Mexican-hat type lateral interactions between neurons in the final layer (discussed further below) induced the organization of the orientation-selective neurons into an orientation map (Fig. 1.14). Using an eigenvector analysis, MacKay and Miller (1990) subsequently demonstrated analytically how the type of receptive field formed in Linsker's model depends on the parameters of the model. Miller *et al.* (1989) applied a similar approach to the formation of ocular dominance columns (Fig. 1.15). Now each output neuron initially receives input from small regions of both the left and right eyes. It was found that the subtractive enforcement of a weight normalization constraint, and assuming zero or negative correlations in activity between the two eyes, was sufficient to drive receptive fields to become strongly dominant for only one eye or the other. Again adding lateral interactions between neurons caused the left- and right-eye dominant receptive fields to become clustered into columnar structures. This model was subsequently extended to the formation of orientation columns (Miller, 1994) and the joint formation of both ocular dominance and orientation columns (Erwin and Miller, 1998).

4.5. Nonlinear Hebbian learning

Learning in the models described above is essentially linear: the weights for all neurons are updated proportionally to their amount of activation. However, the first Hebbian models of the formation of visual maps were actually nonlinear. In particular von der Malsburg (1973) proposed a two-layer model, where the stimuli in the input layer consisted of oriented edges

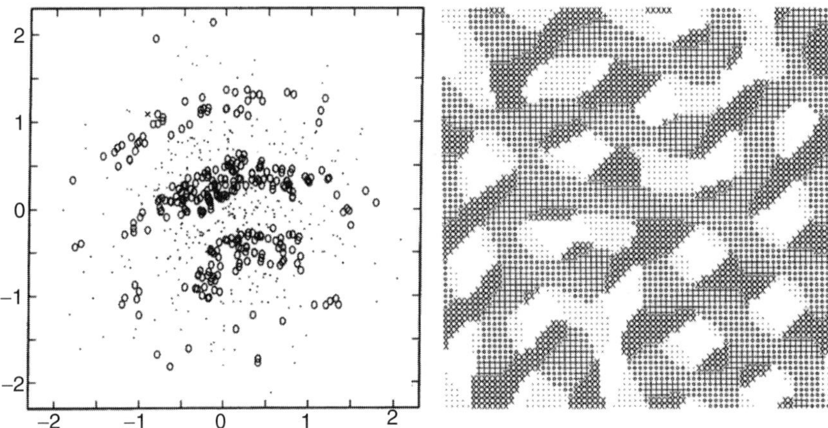

Figure 1.14 Results from the Linsker's (1986) model. Left panel: The receptive field of a V1 neuron generated from Linsker's model. Circles represent negative weights and dots represent positive weights. It can be seen that this cell is selective for both orientation and spatial frequency. Right panel: An orientation preference map generated by Linsker's model. Each different symbol corresponds to a different preferred orientation. Several pinwheels can be seen (from Linsker, 1986).

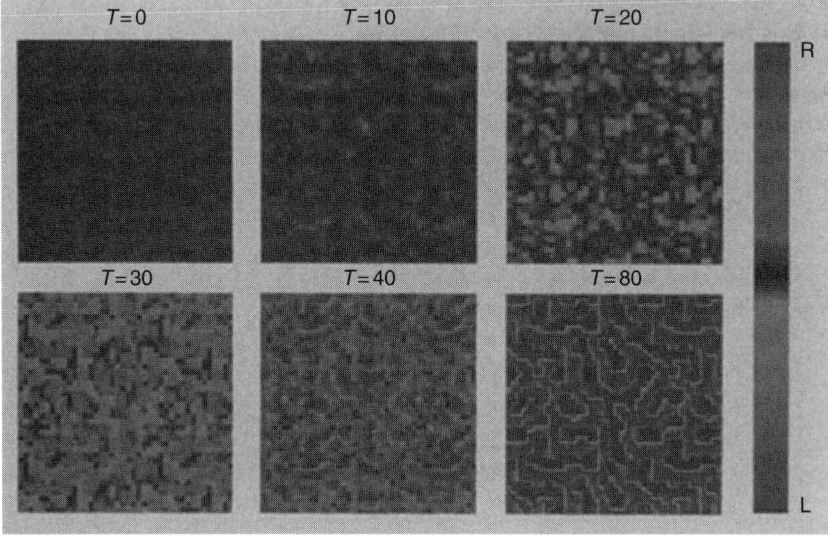

Figure 1.15 Time development of ocular dominance columns in the model of Miller *et al.* (1989). Segregation from an initially unsegregated state can be seen as development proceeds. Developmental time is measured in arbitrary units (from Miller *et al.*, 1989; reproduced with permission from AAAS). (See Color Insert.)

or bars. The output (cortical) layer consisted of both excitatory and inhibitory neurons, with specific patterns of lateral connectivity between them. When an input pattern was presented to the network the input to each cortical neuron was calculated according to a rule similar to Eq. (1.1), but the neuron was only active if its total input exceeded a threshold. The response of each cortical neuron to the input was then iteratively recalculated, taking into account the additional inputs coming from the lateral connections. Once activity in the cortical layer had settled to a stable state the afferent weights were then updated by a Hebbian rule similar to Eq. (1.2). However, the crucial difference with the linear models described above is the nonlinearity of the activation rule: mathematical analysis is now much harder, and in particular the outcome cannot be predicted from the principal components of the correlation matrix of the inputs. However (and remarkably given the computing resources available at the time), von der Malsburg (1973) was able to show by simulation that this model produces oriented receptive fields in the cortical layer, organized in a map-like structure (Fig. 1.16). A very similar approach was then applied to the formation of topography (Willshaw and von der Malsburg, 1976) and ocular dominance columns (von der Malsburg and Willshaw, 1976).

Figure 1.16 Orientation preference map generated from the model of von der Malsburg (1973). This was the first such map produced by a computational model, predating by a decade the detection of such overall map structure experimentally (from von der Malsburg, 1973; reproduced with permission from Springer Science + Business Media).

Another important nonlinear learning rule is that of Bienenstock *et al.* (1982). This has a form similar to that of Eq. (1.2), except that the size of the weight change is also scaled by a factor comparing the current activity of the neuron with a threshold activity θ:

$$\Delta \boldsymbol{w} \propto y\boldsymbol{x}(y - \theta).$$

This allows positive and negative changes in the weights, and adapting θ appropriately allows the learning to be stable. This rule can produce a variety of receptive fields matching those seen in reality, and direct experimental evidence for such a sliding threshold has subsequently been obtained (reviewed in Bear, 2003).

4.6. Competitive learning

In models such as that of von der Malsburg (1973), the presentation of an input pattern leads to the gradual emergence of one or a few hotspots of activity in the output layer, as the activity is iterated through the lateral connections. This iteration process is time consuming, and convergence requires the careful control of several parameters. Kohonen (1982) proposed a drastic but effective simplification of this approach, whereby it is assumed *a priori* that the pattern of activity in the output layer that eventually results from each input pattern will be a "hump" centered on the "winning" neuron; that is, the neuron which received the largest feedforward activation initially. Standard Hebbian learning is then performed, which amounts to adapting the feedforward weights of each neuron as a function of their distance in the output layer from the winning neuron. This is a computationally efficient and highly robust method for forming topographic maps, and has found wide application both in modeling map formation in the brain (discussed below) and in technological applications (Kohonen, 1995).

An alternative way of viewing Kohonen's algorithm is as an extension of another nonlinear approach to Hebbian learning, competitive learning (Rumelhart and Zipser, 1986). In competitive learning, inputs are presented to a layer of output neurons which are assumed to laterally inhibit each other, so that the neuron with the largest initial activity suppresses the activity of all other output neurons. The weights of this "winning" neuron are then updated as usual. This type of algorithm is highly effective for finding clusters in the input data: each input pattern is effectively assigned to one neuron, whose weights then provide the exemplar for that type of input. Barrow (1987) used this approach to model the development of oriented receptive fields in visual cortex. The activity y of each neuron was calculated as in Eq. (1.1), but then only the neuron with the largest y value had its weights updated according to Eq. (1.2). When given fragments of natural images as input patterns, this model produced oriented receptive fields resembling those in V1. Orientation map development using a competitive (Kohonen) approach was modeled by Obermayer *et al.* (1990), and a related competitive model was proposed for the development of ocular dominance columns by Goodhill (1993) (Fig. 1.17).

The competitive approaches described so far are often described as "hard," in the sense of being "winner-take-all." In "soft" competitive learning, *all* neurons in the output layer are updated, by an amount that takes into account both their feedforward activation and the activity of other output neurons. An example is the elastic net, originally developed as a method for solving the *Travelling Salesman Problem* (Durbin and Willshaw, 1987). Here, each input pattern is assumed to have the same amount of efficacy, but this is shared out in a nonlinear way between the output neurons. The elastic net was subsequently shown to be a highly effective tool for modeling the development of ocular dominance (Goodhill and Willshaw, 1990) and orientation (Durbin and Mitchison, 1990) maps in V1. Together with related approaches based on the Kohonen algorithm (Obermayer *et al.*, 1992), these competitive methods have not been surpassed in their ability to predict the fine details of the overall geometric structure of feature maps in V1, for both normal and abnormal development (Carreira-Perpiñán and Goodhill, 2004; Carreira-Perpiñán *et al.*, 2005; Erwin *et al.*, 1995; Farley *et al.*, 2007; Giacomantonio and Goodhill,

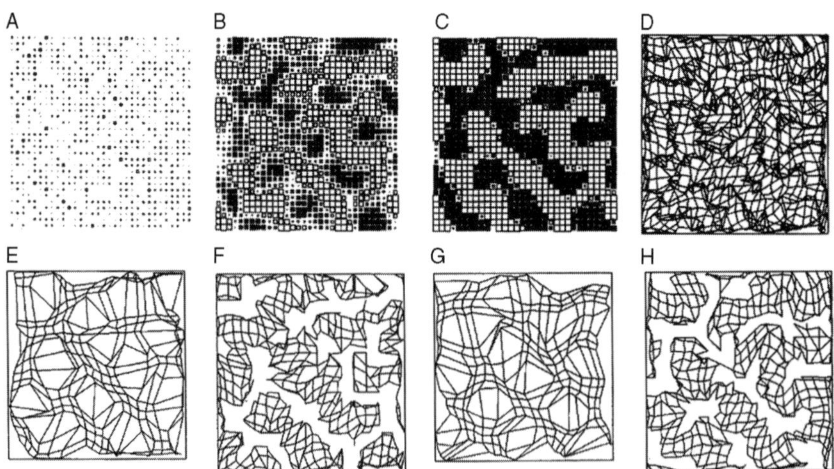

Figure 1.17 Simultaneous formation of ocular dominance columns and a topographic map in the model of Goodhill (1993). (A)–(C) A progression of time points, where the size of each square represents its degree of selectivity. (D)–(H) The center of mass of the weights of each cortical or retinal neuron is represented as a point in the retinal or cortical space, respectively, and neighboring lines are connected to form a grid. (D) Representation of cortical topography for both eyes. (E) and (G) Representation of retinal topography for the right and left eyes, respectively. (F) and (H) Representation of cortical topography for right and left eye, respectively (from Goodhill, 1993; reproduced with permission from Springer Science + Business Media).

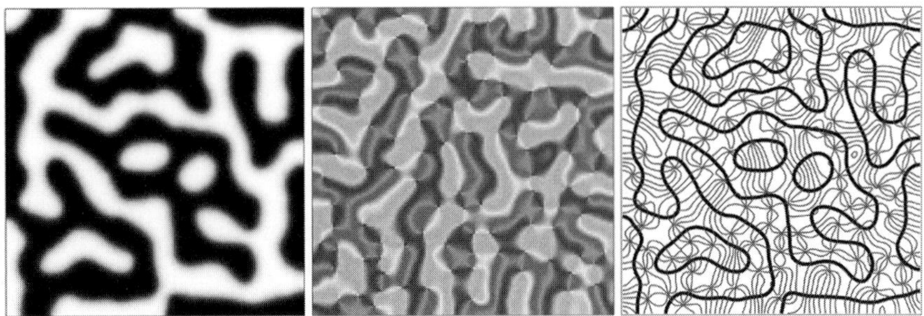

Figure 1.18 Elastic net simulations of visual map development. Left panel: Ocular dominance map. Middle panel: Orientation preference map. Right panel: Superimposed contours of ocular dominance and orientation. Note the tendency to intersect at steep angles, and for pinwheels to lie at the center of ocular dominance columns (see Carreira-Perpiñán *et al.*, 2005 for more details). (See Color Insert.)

2007; Goodhill, 2007; Goodhill and Cimponeriu, 2000; Goodhill *et al.*, 1997; Swindale, 1996; Yu *et al.*, 2005). Figure 1.18 shows ocular dominance and orientation preference maps generated by the elastic net simulations in Carreira-Perpiñán *et al.* (2005).

4.7. Spike-timing-dependent plasticity

All the mathematical instantiations of Hebb's rule we have considered so far were developed prior to the general realization that the relative timing of the presynaptic and postsynaptic spike can crucially affect whether synapses increase or decrease—so-called spike-timing-dependent plasticity (STDP) (Bi and Poo, 1998; Markram *et al.*, 1997; reviewed in Dan and Poo, 2006). While mathematical models that do not address fine-scale timing issues in synaptic plasticity can still be extremely useful for addressing many problems (see for instance the examples above), the inclusion of spike timing opens up a large array of new computational issues to consider (Brette *et al.*, 2007; Burkitt *et al.*, 2004; Kepecs *et al.*, 2002; Morrison *et al.*, 2008).

A relatively simple mathematical model of such a learning rule was proposed by Song *et al.* (2000), which can be expressed in terms of a function $F(t)$ which specifies the amount of increase or decrease in synaptic strength when pre- and postsynaptic spikes occur at times t_{pre} and t_{post}, respectively:

$$\begin{aligned} F(\Delta t) = {} & +A_+ e^{\Delta t/\tau_+}, && \text{if} \quad \Delta t < 0, \\ & -A_- e^{-\Delta t/\tau_-}, && \text{if} \quad \Delta t > 0, \end{aligned}$$

where $\Delta t = t_{\text{pre}} - t_{\text{post}}$, and A_+ and A_- are positive scaling factors. In the completely symmetric case $A_+ = A_-$ and $\tau_+ = \tau_-$. A variety of asymmetries are now known experimentally (Dan and Poo, 2006), many of which can be captured at least roughly by varying the ratios A_+/A_- and τ_+/τ_-. This rule is innately competitive, meaning that the rather unphysiological devices such as weight normalization required in many of the models discussed previously are no longer necessary to achieve competition. It also has the properties of causing postsynaptic neurons to reduce their latency, and tending to make weights go to their maximum or minimum values—a bimodal distribution. (In another version of this rule, where the size of the weight change depends on the current synaptic strength, a unimodal distribution of weights results; discussed further in Morrison *et al.* (2008).) There are also complex issues involved in deciding exactly which postsynaptic spikes the rule applies to; two extremes being all postsynaptic spikes, and just the one closest to the presynaptic spike.

Song and Abbott (2001) used the rule of Song *et al.* (2000) to model cortical map formation, including the development of ocular dominance columns. In this model lateral as well as feedforward weights were adapted by the same STDP rule. A key advantage of this approach over the non-STDP models discussed above is that maps can reach a stable state, but are then still free to change their structure if the input statistics change. In particular, Song and Abbott (2001) showed that "lesioning" some of the inputs in the adult state causes appropriate map rearrangement. Subsequent work by Young *et al.* (2007) showed directly that STDP but not more standard Hebbian learning rules were capable of reproducing a particular type of postlesion plasticity observed in cat visual cortex. Both these authors and Song and Abbott (2001) highlighted the "didactic" properties of STDP learning rules: if neuron A spikes before neuron B and they are reciprocally connected, then STDP tends to cause neuron B to develop similar response properties to neuron A. STDP learning rules have also been applied to, for instance, the development of directionally selective cells in primary visual cortex (Wenisch *et al.*, 2005) (Fig. 1.19).

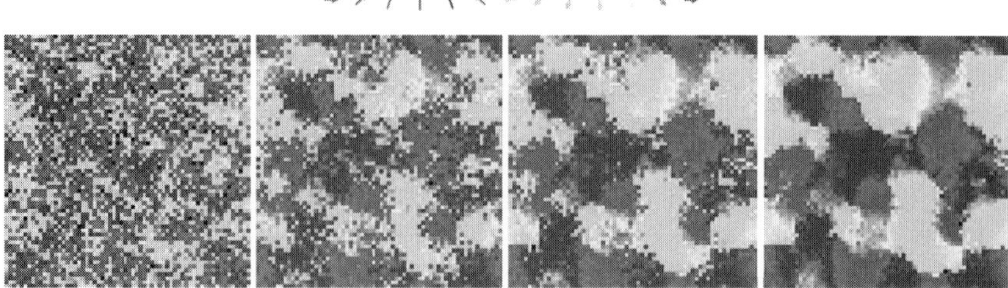

Figure 1.19 Development of a direction preference map in the STDP-based model of Wenisch *et al.* (2005). Sequence of map development at four different time points. Colors corresponding to different direction preferences are shown above the simulations (from Wenisch *et al.*, 2005; reproduced with permission from Springer Science + Business Media). (See Color Insert.)

4.8. Functional modeling

A quite different theoretical approach to neural wiring development is functional modeling. The methods we have discussed up to now generally start with equations intended to capture at least something of the biology of Hebbian synaptic plasticity. By contrast in functional models, the focus is on the underlying computational goal of the developmental process. In particular, functional models generally propose that the purpose of development (at least in certain contexts) is to optimize some information-theoretic measure of the "quality" of the representation (for discussions, see Linsker, 1990; Simoncelli, 2003). Although in some cases the optimization process has a Hebbian interpretation, this is less important for these approaches than a good match between real and simulated receptive fields.

One popular approach in this class is independent component analysis (ICA) (Bell and Sejnowski, 1997; Caywood *et al.*, 2004; Doi *et al.*, 2003; Hsu and Dayan, 2007; Hyvärinen and Hoyer, 2001; Hyvärinen and Köster, 2007; van Hateren and Ruderman, 1998; van Hateren and van der Schaaf, 1998). We saw earlier how a linear Hebbian rule can lead to weight distributions which find the principal components of the input patterns. Rather than maximizing the variance as in PCA, in ICA the goal is to find weight distributions which maximize the statistical independence of the different components. Intuitively, the idea is that sensory inputs (e.g., images) are produced by summing several independent causes (e.g., some basis functions, such as Gabor patches), so that by maximizing independence one can hopefully recover the underlying causes of the input.

A closely related approach, motivated by energetic considerations, is to maximize the sparseness of the representation, that is, minimize the number of neurons that fire in response to any particular input (e.g., Olshausen and Field, 2004). A seminal example of this "sparse-coding" approach is the work of Olshausen and Field (1996). They trained a set of simulated V1 neurons, using a learning rule which maximized response sparseness, on natural images (somewhat analogously to the work of Barrow, 1987 in competitive learning mentioned earlier). They recovered a set of receptive fields that were both orientation selective and localized in space, in particular Gabor-like (see Fig. 1.20). For a much more extensive discussion of these types of approaches to learning see Chapter 10 of Dayan and Abbott (2001).

5. Discussion

Our main conclusions are as follows:

(1) **Axon guidance**. A complete model of axon guidance will ultimately require us to understand the detailed biophysical mechanisms

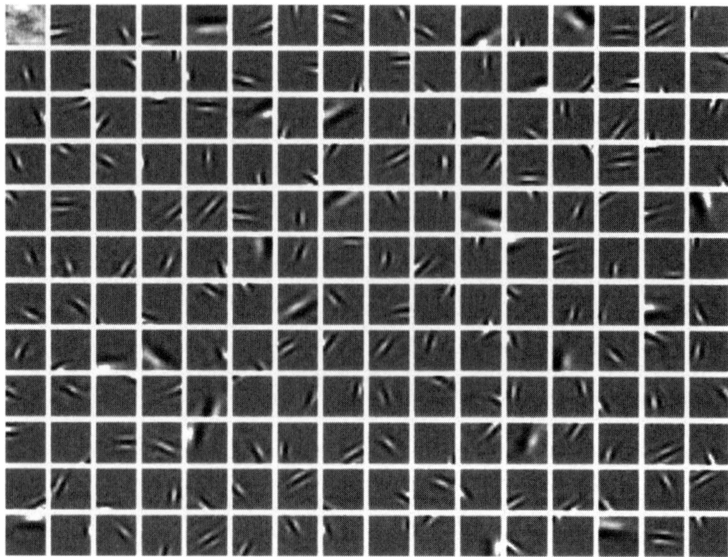

Figure 1.20 Receptive fields of simulated V1 neurons. One hundred and ninety-two basis functions trained on fragments of natural scenes using a sparse-coding algorithm (Olshausen and Field, 1996). These basis functions resemble the receptive fields of neurons in primary visual cortex in that they are localized, oriented, Gabor-like, and display a range of spatial frequency. In this case no topographic structure was imposed on the map (from Olshausen and Field, 1996; reproduced with permission from Macmillan Publishers Ltd.).

underlying growth cone behavior, which will in turn require access to quantitative data to constrain models. Modeling axon guidance is still in its early stages, perhaps because these mechanisms have not yet been well described experimentally (compared with, for example, bacterial chemotaxis), and because quantitative kinetic data is still lacking. Despite these challenges, progress has been made at the broad level of understanding physical constraints faced by growth cones (e.g., in sensing shallow gradients, and filopodial protrusion).

(2) **Retinotectal mapping**. Models in retinotectal mapping have established a number of mechanisms that are capable of producing retinotopic maps, and have described how they can do so. However, in general models have tended to be abstract rather than mechanistic, and perhaps because of this fail to explain important *in vitro* and *in vivo* results. Addressing this should be one focus of future modeling work. Another is to meet challenges presented by new data, which suggest that assumptions regarding known mechanisms (such as competition, chemoaffinity, and activity) and the interactions between them may need to be revisited (Cang *et al.*, 2008a; Gosse *et al.*, 2008; Hua *et al.*, 2005; Nicol *et al.*, 2007).

(3) **Activity-dependent development**. The unfolding of the genetic program alone does not, of course, fully determine nervous system structure. We have seen that by tying theory closely with results from neuroimaging and neurophysiology, a relatively simple set of assumptions about activity in the developing visual system can be used reproduce important features of visual maps seen *in vivo*. Hence by modeling the influence of activity in the development of neural circuitry, we can obtain a quantitative understanding of the influence of the environment on brain wiring. One challenge for future modeling in this field is to use these ideas to tease apart the relative importance of environmentally driven and intrinsic mechanisms for particular problems (e.g., whether certain properties of visual maps are under genetic control or are due to higher order statistics of natural scenes).

We have canvassed some important features of these three fields individually, but how do they relate to each other? We can think of retinotectal mapping as an applied or population version of axon guidance models, in that it should resemble models of individual axon guidance in the limit of small numbers of axons. In turn, activity-dependent processes depend on initial connections being set up by processes such as those in retinotectal mapping. So these areas are certainly related, yet they are generally treated as being quite separate. One reason may be that traditional reductionist modeling usually involves choosing just one particular time and length scale that is representative for the particular problem, and working within that. However, the development of brain wiring, like many biological processes of interest, occurs in regimes where multiple time and length scales interact. As a result, traditional approaches lead to fragmenting of models of these processes, making them appear more distinct than they actually are. For this reason, and almost certainly for many other reasons, biology is difficult to describe in as simple terms as the Ideal Gas Law discussed in the introduction. It may be that unifying principles for biology take on a different form than the laws of physics, so that different approaches may be necessary to solve these problems. For instance, ideal observer analysis (Körding, 2007) and wire minimization approaches (Chklovskii *et al.*, 2002) have proven useful and may drive further progress in understanding neural development. These are examples of normative methods, which seek to find the optimal solution to a given computational task or goal, given certain constraints (i.e., what computational strategy the organism should take to achieve a desired outcome).

There are other challenges for modelers besides the multiscale nature of problems in neural development. Generalizing models of more specific systems (e.g., modeling in the visual system) is not straightforward, but may inform us about brain wiring in general. We have already suggested that a lack of quantitative data is a barrier both to furthering modeling in individual areas, and for marrying together apparently disparate fields. Better communication between experimental and theoretical disciplines will not only facilitate the gathering of this kind of data, but it will also maximize their synergy. There are many reasons why such synergy would be of mutual benefit. Theoretical approaches can allow us to identify hidden assumptions, rapidly generate and test hypotheses, and make accurate predictions; however, modeling is ultimately a solipsistic activity without experimental input and verification. Similarly, the kinds of questions one can ask experimentally are driven strongly by the theoretical framework within which one poses those questions. We suggest that in the future, understanding in neuroscience will not be driven by either field independently, but by thoughtful and respectful collaboration. In this respect, we are motivated by the great success achieved by similar modeling work in eukaryotic and bacterial cell chemotaxis (Paliwal *et al.*, 2007; Tindall *et al.*, 2008a,b), morphogen gradients (Bollenbach *et al.*, 2008; Gregor *et al.*, 2007a,b), gene regulatory networks (Jaeger *et al.*, 2004), and a range of other areas (Mogilner *et al.*, 2006). We look forward to similar success in the field of neural wiring development, as detailed and realistic mathematical models are vital to exploring exciting potential applications in the emerging fields of bioengineering and regenerative medicine; for example, the creation of organic self-wiring computers and bioelectrical interfaces, or guidance and rewiring in cell therapies.

REFERENCES

Abbott, L. F. (2008). Theoretical neuroscience rising. *Neuron* **60,** 489–495.

Ackley, D., Hinton, G., and Sejnowski, T. (1985). A learning algorithm for Boltzmann machines. *Cogn. Sci.* **9,** 147–169.

Aeschlimann, M., and Tettoni, L. (2001). Biophysical model of axonal pathfinding. *Neurocomputing* **38,** 87–92.

Barrow, H. (1987). Learning receptive fields. *Proc. IEEE 1st Annu. Conf. Neural Networks* **IV,** 115–121.

Bear, M. F. (2003). Bidirectional synaptic plasticity: From theory to reality. *Philos. Trans. R. Soc. Lond. B Biol. Sci.* **358,** 649–655.

Bell, A. J., and Sejnowski, T. J. (1997). The "independent components" of natural scenes are edge filters. *Vision Res.* **37,** 3327–3338.

Berg, H. C., and Purcell, E. M. (1977). Physics of chemoreception. *Biophys. J.* **20,** 193–219.

Bi, G. Q., and Poo, M. M. (1998). Synaptic modifications in cultured hippocampal neurons: Dependence on spike timing, synaptic strength, and postsynaptic cell type. *J. Neurosci.* **18,** 10464–10472.

Bienenstock, E. L., Cooper, L. N., and Munro, P. W. (1982). Theory for the development of neuron selectivity: Orientation specificity and binocular interaction in visual cortex. *J. Neurosci.* **2,** 32–48.

Bollenbach, T., Pantazis, P., Kicheva, A., Bökel, C., González-Gaitán, M., and Jülicher, F. (2008). Precision of the Dpp gradient. *Development* **135,** 1137–1146.

Brette, R., Rudolph, M., Carnevale, T., Hines, M., Beeman, D., Bower, J. M., Diesmann, M., Morrison, A., Goodman, P. H., Harris, F. C., Zirpe, M., Natschlgër, T., *et al.* (2007). Simulation of networks of spiking neurons: A review of tools and strategies. *J. Comput. Neurosci.* **23,** 349–398.

Brown, A., Yates, P. A., Burrola, P., Ortuño, D., Vaidya, A., Jessell, T. M., Pfaff, S. L., O'Leary, D. D., and Lemke, G. (2000). Topographic mapping from the retina to the midbrain is controlled by relative but not absolute levels of EphA receptor signaling. *Cell* **102,** 77–88.

Buettner, H. M. (1995). Computer simulation of nerve growth cone filopodial dynamics for visualization and analysis. *Cell Motil. Cytoskel.* **32,** 187–204.

Buettner, H. (1996). Analysis of cell-target encounter by random filopodial projections. *AIChE J.* **42,** 1127–1138.

Buettner, H. M., Pittman, R. N., and Ivins, J. K. (1994). A model of neurite extension across regions of nonpermissive substrate: Simulations based on experimental measurement of growth cone motility and filopodial dynamics. *Dev. Biol.* **163,** 407–422.

Burkitt, A. N., Meffin, H., and Grayden, D. B. (2004). Spike-timing-dependent plasticity: The relationship to rate-based learning for models with weight dynamics determined by a stable fixed point. *Neural Comput.* **16,** 885–940.

Cang, J., Niell, C. M., Liu, X., Pfeiffenberger, C., Feldheim, D. A., and Stryker, M. P. (2008a). Selective disruption of one Cartesian axis of cortical maps and receptive fields by deficiency in ephrin-As and structured activity. *Neuron* **57,** 511–523.

Cang, J., Wang, L., Stryker, M. P., and Feldheim, D. A. (2008b). Roles of ephrin-As and structured activity in the development of functional maps in the superior colliculus. *J. Neurosci.* **28,** 11015–11023.

Carreira-Perpiñán, M., and Goodhill, G. (2004). Influence of lateral connections on the structure of cortical maps. *J. Neurophysiol.* **92,** 2947–2959.

Carreira-Perpiñán, M., Lister, R., and Goodhill, G. (2005). A computational model for the development of multiple maps in primary visual cortex. *Cereb. Cortex* **15,** 1222–1233.

Caywood, M. S., Willmore, B., and Tolhurst, D. J. (2004). Independent components of color natural scenes resemble V1 neurons in their spatial and color tuning. *J. Neurophysiol.* **91,** 2859–2873.

Chklovskii, D. B., Schikorski, T., and Stevens, C. F. (2002). Wiring optimization in cortical circuits. *Neuron* **34,** 341–347.

Cowan, J., and Friedman, A. (1990). Development and regeneration of eye–brain maps: A computational model. *In* "Advances in Neural Information Processing Systems II" (D.S. Touretzky, Ed.), pp. 92–99. Morgan Kaufman, San Mateo, CA.

Cowan, J., and Friedman, A. (1991). Studies of a model for the development and regeneration of eye–brain maps. *In* "Advances in Neural Information Processing Systems III" (D.S. Touretzky, Ed.), pp. 3–10. Morgan Kaufman, San Mateo, CA.

Dan, Y., and Poo, M. M. (2006). Spike timing-dependent plasticity: From synapse to perception. *Physiol. Rev.* **86,** 1033–1048.

Dayan, P., and Abbott, L. (2001). "Theoretical Neuroscience" MIT Press, Cambridge, MA.

Debski, E. A., and Cline, H. T. (2002). Activity-dependent mapping in the retinotectal projection. *Curr. Opin. Neurobiol.* **12,** 93–99.

Doi, E., Inui, T., Lee, T. W., Wachtler, T., and Sejnowski, T. J. (2003). Spatiochromatic receptive field properties derived from information-theoretic analyses of cone mosaic responses to natural scenes. *Neural Comput.* **15,** 397–417.

Durbin, R., and Mitchison, G. (1990). A dimension reduction framework for understanding cortical maps. *Nature* **343,** 644–647.

Durbin, R., and Willshaw, D. (1987). An analogue approach to the travelling salesman problem using an elastic net method. *Nature* **326,** 689–691.

Erwin, E., and Miller, K. D. (1998). Correlation-based development of ocularly matched orientation and ocular dominance maps: Determination of required input activities. *J. Neurosci.* **18,** 9870–9895.

Erwin, E., Obermayer, K., and Schulten, K. (1995). Models of orientation and ocular dominance columns in the visual cortex: A critical comparison. *Neural Comput.* **7,** 425–468.

Farley, B. J., Yu, H., Jin, D. Z., and Sur, M. (2007). Alteration of visual input results in a coordinated reorganization of multiple visual cortex maps. *J. Neurosci.* **27,** 10299–10310.

Feldheim, D. A., Kim, Y. I., Bergemann, A. D., Frisén, J., Barbacid, M., and Flanagan, J. G. (2000). Genetic analysis of ephrin-A2 and ephrin-A5 shows their requirement in multiple aspects of retinocollicular mapping. *Neuron* **25,** 563–574.

Feng, J., Pan, H., and Roychowdhury, V. P. (1997). A rigorous analysis of Linsker's unsupervised Hebbian learning. *Neural Networks* **10,** 705–720.

Flanagan, J. G., and Vanderhaeghen, P. (1998). The ephrins and Eph receptors in neural development. *Annu. Rev. Neurosci.* **21,** 309–345.

Fraser, S. E. (1980). Differential adhesion approach to the patterning of nerve connections. *Dev. Biol.* **79,** 453–464.

Fraser, S. E., and Perkel, D. H. (1990). Competitive and positional cues in the patterning of nerve connections. *J. Neurobiol.* **21,** 51–72.

Fujisawa, H., Tani, N., Watanabe, K., and Ibata, Y. (1982). Branching of regenerating retinal axons and preferential selection of appropriate branches for specific neuronal connection in the newt. *Dev. Biol.* **90,** 43–57.

Gaze, R. M., and Keating, M. J. (1972). The visual system and "neuronal specificity" *Nature* **237,** 375–378.

Gaze, R. M., Keating, M. J., and Chung, S. H. (1974). The evolution of the retinotectal map during development in Xenopus. *Proc. R. Soc. Lond. B Biol. Sci.* **185,** 301–330.

Giacomantonio, C. E., and Goodhill, G. J. (2007). The effect of angioscotomas on map structure in primary visual cortex. *J. Neurosci.* **27,** 4935–4946.

Gierer, A. (1981). Development of projections between areas of the nervous system. *Biol. Cybern.* **42,** 69–78.

Gierer, A. (1983). Model for the retino-tectal projection. *Proc. R. Soc. Lond. B Biol. Sci.* **218,** 77–93.

Gierer, A. (1987). Directional cues for growing axons forming the retinotectal projection. *Development* **101**(3), 479–489.

Giniger, E. (2002). How do Rho family GTPases direct axon growth and guidance? A proposal relating signaling pathways to growth cone mechanics. *Differentiation* **70,** 385–396.

Goodhill, G. J. (1993). Topography and ocular dominance: A model exploring positive correlations. *Biol. Cybern.* **69,** 109–118.

Goodhill, G. J. (1997). Diffusion in axon guidance. *Eur. J. Neurosci.* **9,** 1414–1421.

Goodhill, G. J. (1998). Gradients for retinotectal mapping. *In* "Advances in Neural Information Processing Systems" (M. I. Jordan, M. J. Kearns, and S. A. Solla, Eds.), vol. 10, pp. 152–158. MIT Press, Cambridge, MA.

Goodhill, G. J. (2000). Dating behavior of the retinal ganglion cell. *Neuron* **25,** 501–503.

Goodhill, G. J. (2007). Contributions of theoretical modeling to the understanding of neural map development. *Neuron* **56,** 301–311.

Goodhill, G. J., and Baier, H. (1998). Axon guidance: Stretching gradients to the limit. *Neural Comput.* **10,** 521–527.

Goodhill, G., and Barrow, H. (1994). The role of weight normalization in competitive learning. *Neural Comput.* **6,** 255–269.

Goodhill, G. J., and Cimponeriu, A. (2000). Analysis of the elastic net model applied to the formation of ocular dominance and orientation columns. *Network* **11,** 153–168.

Goodhill, G. J., and Urbach, J. S. (1999). Theoretical analysis of gradient detection by growth cones. *J. Neurobiol.* **41,** 230–241.
Goodhill, G., and Willshaw, D. (1990). Application of the elastic net algorithm to the formation of ocular dominance stripes. *Network* **1,** 41–59.
Goodhill, G. J., and Xu, J. (2005). The development of retinotectal maps: A review of models based on molecular gradients. *Network* **16,** 5–34.
Goodhill, G. J., Bates, K. R., and Montague, P. R. (1997). Influences on the global structure of cortical maps. *Proc. R. Soc. Lond. B Biol. Sci.* **264,** 649–655.
Goodhill, G. J., Gu, M., and Urbach, J. S. (2004). Predicting axonal response to molecular gradients with a computational model of filopodial dynamics. *Neural Comput.* **16,** 2221–2243.
Gosse, N. J., Nevin, L. M., and Baier, H. (2008). Retinotopic order in the absence of axon competition. *Nature* **452,** 892–895.
Graham, B. P., and van Ooyen, A. (2001). Compartmental models of growing neurites. *Neurocomputing* **38–40,** 31–36.
Gregor, T., Wieschaus, E. F., McGregor, A. P., Bialek, W., and Tank, D. W. (2007a). Stability and nuclear dynamics of the Bicoid morphogen gradient. *Cell* **130,** 141–152.
Gregor, T., Tank, D. W., Wieschaus, E. F., and Bialek, W. (2007b). Probing the limits to positional information. *Cell* **130,** 153–164.
Halloran, M. C., and Wolman, M. A. (2006). Repulsion or adhesion: Receptors make the call. *Curr. Opin. Cell Biol.* **18,** 533–540.
Hansen, M. J., Dallal, G. E., and Flanagan, J. G. (2004). Retinal axon response to ephrin-As shows a graded, concentration-dependent transition from growth promotion to inhibition. *Neuron* **42,** 717–730.
Hebb, D. O. (1949). "The Organization of Behaviour." Wiley, New York.
Hely, T. A., and Willshaw, D. J. (1998). Short-term interactions between microtubules and actin filaments underlie long-term behaviour in neuronal growth cones. *Proc. Biol. Sci.* **265,** 1801–1807.
Hentschel, H. G., and van Ooyen, A. (1999). Models of axon guidance and bundling during development. *Proc. Biol. Sci.* **266,** 2231–2238.
Hertz, J., Krogh, A., and Palmer, R. (1991). Introduction to the theory of neural computation. *In* "Lecture Notes in the Santa Fe Institute Studies in the Sciences of Complexity" Addison-Wesley, Redwood City, CA.
Hinton, G. (1989). Connectionist learning procedures. *Artif. Intell.* **40,** 185–234.
Hinton, G. E. (1992). How neural networks learn from experience. *Sci. Am.* **267,** 144–151.
Hinton, G. E. (2007). Learning multiple layers of representation. *Trends Cogn. Sci.* **11,** 428–434.
Holt, C. E. (1984). Does timing of axon outgrowth influence initial retinotectal topography in Xenopus? *J. Neurosci.* **4,** 1130–1152.
Honda, H. (1998). Topographic mapping in the retinotectal projection by means of complementary ligand and receptor gradients: A computer simulation study. *J. Theor. Biol.* **192,** 235–246.
Honda, H. (2003). Competition between retinal ganglion axons for targets under the servomechanism model explains abnormal retinocollicular projection of Eph receptor-overexpressing or ephrin-lacking mice. *J. Neurosci.* **23,** 10368–10377.
Honda, H. (2004). Competitive interactions between retinal ganglion axons for tectal targets explain plasticity of retinotectal projection in the servomechanism model of retinotectal mapping. *Dev. Growth Differ.* **46,** 425–437.
Hope, R. A., Hammond, B. J., and Gaze, R. M. (1976). The arrow model: Retinotectal specificity and map formation in the goldfish visual system. *Proc. R. Soc. Lond. B Biol. Sci.* **194,** 447–466.

Hornberger, M. R., Dtting, D., Ciossek, T., Yamada, T., Handwerker, C., Lang, S., Weth, F., Huf, J., Wessel, R., Logan, C., Tanaka, H., and Drescher, U. (1999). Modulation of EphA receptor function by coexpressed ephrinA ligands on retinal ganglion cell axons. *Neuron* **22,** 731–742.

Hsu, A. S., and Dayan, P. (2007). An unsupervised learning model of neural plasticity: Orientation selectivity in goggle-reared kittens. *Vision Res.* **47,** 2868–2877.

Hua, J. Y., Smear, M. C., Baier, H., and Smith, S. J. (2005). Regulation of axon growth *in vivo* by activity-based competition. *Nature* **434,** 1022–1026.

Huynh-Do, U., Stein, E., Lane, A. A., Liu, H., Cerretti, D. P., and Daniel, T. O. (1999). Surface densities of ephrin-B1 determine EphB1-coupled activation of cell attachment through $\alpha_v\beta_3$ and $\alpha_5\beta_1$ integrins. *EMBO J.* **18,** 2165–2173.

Hyvärinen, A., and Hoyer, P. O. (2001). A two-layer sparse coding model learns simple and complex cell receptive fields and topography from natural images. *Vision Res.* **41,** 2413–2423.

Hyvärinen, A., and Köster, U. (2007). Complex cell pooling and the statistics of natural images. *Network* **18,** 81–100.

Jaeger, J., Surkova, S., Blagov, M., Janssens, H., Kosman, D., Kozlov, K. N., Manu, E., Myasnikova, E., Vanario-Alonso, C. E., Samsonova, M., Sharp, D. H., and Reinitz, J. (2004). Dynamic control of positional information in the early drosophila embryo. *Nature* **430,** 368–371.

Jilkine, A., Marée, A. F. M., and Edelstein-Keshet, L. (2007). Mathematical model for spatial segregation of the Rho-family GTPases based on inhibitory crosstalk. *Bull. Math. Biol.* **69,** 1943–1978.

Kaethner, R. J., and Stuermer, C. A. (1992). Dynamics of terminal arbor formation and target approach of retinotectal axons in living zebrafish embryos: A time-lapse study of single axons. *J. Neurosci.* **12,** 3257–3271.

Katz, M. J., and Lasek, R. J. (1985). Early axon patterns of the spinal cord: Experiments with a computer. *Dev. Biol.* **109,** 140–149.

Katz, L. C., and Shatz, C. J. (1996). Synaptic activity and the construction of cortical circuits. *Science* **274,** 1133–1138.

Katz, M. J., George, E. B., and Gilbert, L. J. (1984). Axonal elongation as a stochastic walk. *Cell Motil.* **4,** 351–370.

Kepecs, A., van Rossum, M. C. W., Song, S., and Tegner, J. (2002). Spike-timing-dependent plasticity: Common themes and divergent vistas. *Biol. Cybern.* **87,** 446–458.

Kiddie, G., McLean, D., Ooyen, A. V., and Graham, B. (2005). Biologically plausible models of neurite outgrowth. *Prog. Brain Res.* **147,** 67–80.

King, C. E., Wallace, A., Rodger, J., Bartlett, C., Beazley, L. D., and Dunlop, S. A. (2003). Transient up-regulation of retinal EphA3 and EphA5, but not ephrin-A2, coincides with re-establishment of a topographic map during optic nerve regeneration in goldfish. *Exp. Neurol.* **183,** 593–599.

Kobayashi, N., and Mundel, P. (1998). A role of microtubules during the formation of cell processes in neuronal and non-neuronal cells. *Cell Tissue Res.* **291,** 163–174.

Kohonen, T. (1982). Self-organized formation of topologically correct feature maps. *Biol. Cybern.* **43,** 59–69.

Kohonen, T. (1995). "Self-Organizing Maps." Springer, Berlin.

Körding, K. (2007). Decision theory: What "should" the nervous system do? *Science* **318,** 606–610.

Koulakov, A. A., and Tsigankov, D. N. (2004). A stochastic model for retinocollicular map development. *BMC Neurosci.* **5,** 30.

Krottje, J. K., and van Ooyen, A. (2007). A mathematical framework for modeling axon guidance. *Bull. Math. Biol.* **69,** 3–31.

Krzanowski, W. (1988). "Principles of Multivariate Analysis: A User's Perspective." Oxford Statistical Science Series V3. Oxford University Press, New York.

Kullander, K., and Klein, R. (2002). Mechanisms and functions of Eph and ephrin signalling. *Nat. Rev. Mol. Cell Biol.* **3,** 475–486.

Linsker, R. (1986). From basic network principles to neural architecture (series). *Proc. Natl. Acad. Sci. USA* **83,** 7508–7512, 8390–8394, 8779–8783.

Linsker, R. (1990). Perceptual neural organization: Some approaches based on network models and information theory. *Annu. Rev. Neurosci.* **13,** 257–281.

Löschinger, J., Weth, F., and Bonhoeffer, F. (2000). Reading of concentration gradients by axonal growth cones. *Philos. Trans. R. Soc. Lond. B Biol. Sci.* **355,** 971–982.

MacKay, D., and Miller, K. (1990). Analysis of Linsker's application of Hebbian rules to linear networks. *Network* **1,** 257–297.

Malenka, R. C., and Bear, M. F. (2004). LTP and LTD: An embarrassment of riches. *Neuron* **44,** 5–21.

Markram, H., Lübke, J., Frotscher, M., and Sakmann, B. (1997). Regulation of synaptic efficacy by coincidence of postsynaptic APs and EPSPs. *Science* **275,** 213–215.

Maskery, S., and Shinbrot, T. (2005). Deterministic and stochastic elements of axonal guidance. *Annu. Rev. Biomed. Eng.* **7,** 187–221.

McLaughlin, T., and O'Leary, D. D. M. (2005). Molecular gradients and development of retinotopic maps. *Annu. Rev. Neurosci.* **28,** 327–355.

McLaughlin, T., Torborg, C. L., Feller, M. B., and O'Leary, D. D. M. (2003). Retinotopic map refinement requires spontaneous retinal waves during a brief critical period of development. *Neuron* **40,** 1147–1160.

McLean, D., van Ooyen, A., and Graham, B. P. (2004). Continuum model for tubulin-driven neurite elongation. *Neurocomputing* **58–60,** 511–516.

Meinhardt, H. (1999). Orientation of chemotactic cells and growth cones: Models and mechanisms. *J. Cell Sci.* **112**(Pt. 17), 2867–2874.

Miller, K. D. (1994). A model for the development of simple cell receptive fields and the ordered arrangement of orientation columns through activity-dependent competition between ON- and OFF-center inputs. *J. Neurosci.* **14,** 409–441.

Miller, K. D. (1996). Synaptic economics: Competition and cooperation in synaptic plasticity. *Neuron* **17,** 371–374.

Miller, K., and MacKay, D. (1994). The role of constraints in Hebbian learning. *Neural Comput.* **6,** 100–126.

Miller, K. D., Keller, J. B., and Stryker, M. P. (1989). Ocular dominance column development: Analysis and simulation. *Science* **245,** 605–615.

Ming, G. I., Wong, S. T., Henley, J., Yuan, X. B., Song, H. J., Spitzer, N. C., and Poo, M. M. (2002). Adaptation in the chemotactic guidance of nerve growth cones. *Nature* **417,** 411–418.

Minsky, M., and Papert, S. (1969). "Perceptrons." MIT Press, Cambridge, MA.

Mogilner, A., and Rubinstein, B. (2005). The physics of filopodial protrusion. *Biophys. J.* **89,** 782–795.

Mogilner, A., Wollman, R., and Marshall, W. F. (2006). Quantitative modeling in cell biology: What is it good for? *Dev. Cell* **11,** 279–287.

Morrison, A., Diesmann, M., and Gerstner, W. (2008). Phenomenological models of synaptic plasticity based on spike timing. *Biol. Cybern.* **98,** 459–478.

Mortimer, D., Fothergill, T., Pujic, Z., Richards, L. J., and Goodhill, G. J. (2008). Growth cone chemotaxis. *Trends Neurosci.* **31,** 90–98.

Nakamoto, M., Cheng, H. J., Friedman, G. C., McLaughlin, T., Hansen, M. J., Yoon, C. H., O'Leary, D. D., and Flanagan, J. G. (1996). Topographically specific effects of ELF-1 on retinal axon guidance *in vitro* and retinal axon mapping *in vivo*. *Cell* **86,** 755–766.

Nicol, X., Voyatzis, S., Muzerelle, A., Narboux-Nême, N., Südhof, T. C., Miles, R., and Gaspar, P. (2007). cAMP oscillations and retinal activity are permissive for ephrin signaling during the establishment of the retinotopic map. *Nat. Neurosci.* **10,** 340–347.

Obermayer, K., Ritter, H., and Schulten, K. (1990). A principle for the formation of the spatial structure of cortical feature maps. *Proc. Natl. Acad. Sci. USA* **87,** 8345–8349.

Obermayer, K., Blasdel, C. G., and Schulten, K. (1992). Statistical–mechanical analysis of self-organization and pattern formation during the development of visual maps. *Phys. Rev. A* **45,** 7568–7589.

Odde, D. J., and Buettner, H. M. (1998). Autocorrelation function and power spectrum of two-state random processes used in neurite guidance. *Biophys. J.* **75,** 1189–1196.

Odde, D., Tanaka, E., Hawkins, S., and Buettner, H. (1996). Stochastic dynamics of the nerve growth cone and its microtubules during neurite outgrowth. *Biotechnol. Bioeng.* **50**(4), 452–461.

Oja, E. (1982). A simplified neuron model as a principal component analyzer. *J. Math. Biol.* **15,** 267–273.

Oja, E. (1989). Neural networks, principal components, and subspaces. *Int. J. Neural Syst.* **1,** 61–68.

Olshausen, B. A., and Field, D. J. (1996). Emergence of simple-cell receptive field properties by learning a sparse code for natural images. *Nature* **381,** 607–609.

Olshausen, B. A., and Field, D. J. (2004). Sparse coding of sensory inputs. *Curr. Opin. Neurobiol.* **14,** 481–487.

O'Rourke, N. A., and Fraser, S. E. (1990). Dynamic changes in optic fiber terminal arbors lead to retinotopic map formation: An *in vivo* confocal microscopic study. *Neuron* **5,** 159–171.

O'Rourke, N. A., Cline, H. T., and Fraser, S. E. (1994). Rapid remodeling of retinal arbors in the tectum with and without blockade of synaptic transmission. *Neuron* **12,** 921–934.

Overton, K. J., and Arbib, M. A. (1982a). Systems matching and topographic maps: The branch-arrow model (BAM). *In* "Competition and Cooperation in Neural Nets", Lecture Notes in Biomathematics (S. Amari and M.A. Arbib, Eds.), vol. 45, pp. 202–225. Springer, Berlin.

Overton, K. J., and Arbib, M. A. (1982b). The extended branch-arrow model of the formation of retino-tectal connections. *Biol. Cybern.* **45,** 157–175.

Paliwal, S., Iglesias, P. A., Campbell, K., Hilioti, Z., Groisman, A., and Levchenko, A. (2007). MAPK-mediated bimodal gene expression and adaptive gradient sensing in yeast. *Nature* **446,** 46–51.

Pasquale, E. B. (2005). Eph receptor signalling casts a wide net on cell behaviour. *Nat. Rev. Mol. Cell Biol.* **6,** 462–475.

Piper, M., Salih, S., Weinl, C., Holt, C. E., and Harris, W. A. (2005). Endocytosis-dependent desensitization and protein synthesis-dependent resensitization in retinal growth cone adaptation. *Nat. Neurosci.* **8,** 179–186.

Poliakov, A., Cotrina, M., and Wilkinson, D. G. (2004). Diverse roles of Eph receptors and ephrins in the regulation of cell migration and tissue assembly. *Dev. Cell* **7,** 465–480.

Prestige, M. C., and Willshaw, D. J. (1975). On a role for competition in the formation of patterned neural connexions. *Proc. R. Soc. Lond. B Biol. Sci.* **190,** 77–98.

Rashid, T., Upton, A. L., Blentic, A., Ciossek, T., Knöll, B., Thompson, I. D., and Drescher, U. (2005). Opposing gradients of ephrin-As and EphA7 in the superior colliculus are essential for topographic mapping in the mammalian visual system. *Neuron* **47,** 57–69.

Reber, M., Burrola, P., and Lemke, G. (2004). A relative signalling model for the formation of a topographic neural map. *Nature* **431,** 847–853.

Rodger, J., Bartlett, C. A., Beazley, L. D., and Dunlop, S. A. (2000). Transient up-regulation of the rostrocaudal gradient of ephrin A2 in the tectum coincides with reestablishment of orderly projections during optic nerve regeneration in goldfish. *Exp. Neurol.* **166,** 196–200.

Roskies, A. L., and O'Leary, D. D. (1994). Control of topographic retinal axon branching by inhibitory membrane-bound molecules. *Science* **265,** 799–803.

Rosoff, W. J., Urbach, J. S., Esrick, M. A., McAllister, R. G., Richards, L. J., and Goodhill, G. J. (2004). A new chemotaxis assay shows the extreme sensitivity of axons to molecular gradients. *Nat. Neurosci.* **7,** 678–682.

Rumelhart, D., and Zipser, D. (1986). Feature discovery by competitive learning. *In* "Parallel Distributed Processing: Explorations in the Microstructure of Cognition" (J. L. McClelland and D. E. Rumelhart, Eds.), pp. 151–193. MIT Press, Cambridge, MA.

Ruthazer, E. S., and Cline, H. T. (2004). Insights into activity-dependent map formation from the retinotectal system: A middle-of-the-brain perspective. *J. Neurobiol.* **59,** 134–146.

Ruthazer, E. S., Akerman, C. J., and Cline, H. T. (2003). Control of axon branch dynamics by correlated activity *in vivo*. *Science* **301,** 66–70.

Sakumura, Y., Tsukada, Y., Yamamoto, N., and Ishii, S. (2005). A molecular model for axon guidance based on cross talk between Rho GTPases. *Biophys. J.* **89,** 812–822.

Schmidt, J. T. (1978). Retinal fibers alter tectal positional markers during the expansion of the retinal projection in goldfish. *J. Comp. Neurol.* **177,** 279–295.

Schmidt, J. T. (1990). Long-term potentiation and activity-dependent retinotopic sharpening in the regenerating retinotectal projection of goldfish: Common sensitive period and sensitivity to NMDA blockers. *J. Neurosci.* **10,** 233–246.

Schmidt, J. T., and Easter, S. S. (1978). Independent biaxial reorganization of the retinotectal projection: A reassessment. *Exp. Brain Res.* **31,** 155–162.

Schmidt, J. T., Cicerone, C. M., and Easter, S. S. (1978). Expansion of the half retinal projection to the tectum in goldfish: An electrophysiological and anatomical study. *J. Comp. Neurol.* **177,** 257–277.

Schultz, W., Dayan, P., and Montague, P. R. (1997). A neural substrate of prediction and reward. *Science* **275,** 1593–1599.

Sejnowski, T. J. (1977). Statistical constraints on synaptic plasticity. *J. Theor. Biol.* **69,** 385–389.

Sharma, S. C. (1972). Reformation of retinotectal projections after various tectal ablations in adult goldfish. *Exp. Neurol.* **34,** 171–182.

Simon, D. K., and O'Leary, D. D. (1992). Development of topographic order in the mammalian retinocollicular projection. *J. Neurosci.* **12,** 1212–1232.

Simoncelli, E. P. (2003). Vision and the statistics of the visual environment. *Curr. Opin. Neurobiol.* **13,** 144–149.

Song, S., and Abbott, L. F. (2001). Cortical development and remapping through spike timing-dependent plasticity. *Neuron* **32,** 339–350.

Song, S., Miller, K. D., and Abbott, L. F. (2000). Competitive Hebbian learning through spike-timing-dependent synaptic plasticity. *Nat. Neurosci.* **3,** 919–926.

Sperry, R. (1963). Chemoaffinity in the orderly growth of nerve fiber patterns and connections. *Proc. Natl. Acad. Sci. USA* **50,** 703–710.

Stuermer, C. A. (1986). Pathways of regenerated retinotectal axons in goldfish. I. Optic nerve, tract and tectal fascicle layer. *J. Embryol. Exp. Morphol.* **93,** 1–28.

Sutton, R. (1988). Learning to predict by the methods of temporal differences. *Mach. Learn.* **3,** 9–44.

Swindale, N. V. (1996). The development of topography in the visual cortex: A review of models. *Network* **7,** 161–247.

Thivierge, J. P., and Balaban, E. (2007). Getting into shape: Optimal ligand gradients for axonal guidance. *Biosystems* **90,** 61–77.

Tindall, M. J., Porter, S. L., Maini, P. K., Gaglia, G., and Armitage, J. P. (2008a). Overview of mathematical approaches used to model bacterial chemotaxis I: The single cell. *Bull. Math. Biol.* **70,** 1525–1569.

Tindall, M. J., Maini, P. K., Porter, S. L., and Armitage, J. P. (2008b). Overview of mathematical approaches used to model bacterial chemotaxis II: Bacterial populations. *Bull. Math. Biol.* **70,** 1570–1607.

Tsigankov, D. N., and Koulakov, A. A. (2006). A unifying model for activity-dependent and activity-independent mechanisms predicts complete structure of topographic maps in ephrin-A deficient mice. *J. Comput. Neurosci.* **21,** 101–114.

Udin, S. B., and Fawcett, J. W. (1988). Formation of topographic maps. *Annu. Rev. Neurosci.* **11,** 289–327.

Urbach, J., and Goodhill, G. (1999). Limitations on detection of gradients of diffusible chemicals by axons. *Neurocomputing* **26–27,** 39–43.

van Hateren, J. H., and Ruderman, D. L. (1998). Independent component analysis of natural image sequences yields spatio-temporal filters similar to simple cells in primary visual cortex. *Proc. R. Soc. Lond. B Biol. Sci.* **265,** 2315–2320.

van Hateren, J. H., and van der Schaaf, A. (1998). Independent component filters of natural images compared with simple cells in primary visual cortex. *Proc. R. Soc. Lond. B Biol. Sci.* **265,** 359–366.

van Ooyen, A. (2001). Competition in the development of nerve connections: A review of models. *Network* **12,** R1–R47.

van Ooyen, A. (Ed.), (2003). "Modeling Neural Development." MIT Press, Cambridge, MA.

van Veen, M. P., and van Pelt, J. (1994). Neuritic growth rate described by modeling microtubule dynamics. *Bull. Math. Biol.* **56,** 249–273.

von der Malsburg, C. (1973). Self-organization of orientation sensitive cells in the striate cortex. *Kybernetik* **14,** 85–100.

von der Malsburg, C., and Willshaw, D. (1976). A mechanism for producing continuous neural mappings: Ocularity dominance stripes and ordered retino-tectal projections. *Exp. Brain Res.* Suppl. **1,** 463–469.

von der Malsburg, C., and Willshaw, D. J. (1977). How to label nerve cells so that they can interconnect in an ordered fashion. *Proc. Natl. Acad. Sci. USA* **74,** 5176–5178.

von Philipsborn, A., and Bastmeyer, M. (2007). Mechanisms of gradient detection: A comparison of axon pathfinding with eukaryotic cell migration. *Int. Rev. Cytol.* **263,** 1–62.

Walter, J., Kern-Veits, B., Huf, J., Stolze, B., and Bonhoeffer, F. (1987a). Recognition of position-specific properties of tectal cell membranes by retinal axons *in vitro*. *Development* **101,** 685–696.

Walter, J., Henke-Fahle, S., and Bonhoeffer, F. (1987b). Avoidance of posterior tectal membranes by temporal retinal axons. *Development* **101,** 909–913.

Weber, C., Ritter, H., Cowan, J., and Klaus Obermayer, K. (1997). Development and regeneration of the retinotectal map in goldfish: A computational study. *Philos. Trans. R. Soc. Lond. B Biol. Sci.* **352**(1361), 1603–1623.

Wenisch, O. G., Noll, J., and van Hemmen, J. L. (2005). Spontaneously emerging direction selectivity maps in visual cortex through STDP. *Biol. Cybern.* **93,** 239–247.

Whitelaw, V. A., and Cowan, J. D. (1981). Specificity and plasticity of retinotectal connections: A computational model. *J. Neurosci.* **1,** 1369–1387.

Wilkinson, D. G. (2000). Topographic mapping: Organising by repulsion and competition? *Curr. Biol.* **10,** R447–R451.

Wilkinson, D. G. (2001). Multiple roles of Eph receptors and ephrins in neural development. *Nat. Rev. Neurosci.* **2,** 155–164.

Willshaw, D. (2006). Analysis of mouse EphA knockins and knockouts suggests that retinal axons programme target cells to form ordered retinotopic maps. *Development* **133,** 2705–2717.

Willshaw, D. J., and von der Malsburg, C. (1976). How patterned neural connections can be set up by self-organization. *Proc. R. Soc. Lond. B Biol. Sci.* **194,** 431–445.

Willshaw, D. J., and von der Malsburg, C. (1979). A marker induction mechanism for the establishment of ordered neural mappings: Its application to the retinotectal problem. *Philos. Trans. R. Soc. Lond. B Biol. Sci.* **287,** 203–243.

Xu, J., Rosoff, W. J., Urbach, J. S., and Goodhill, G. J. (2005). Adaptation is not required to explain the long-term response of axons to molecular gradients. *Development* **132,** 4545–4552.

Yates, P. A., Roskies, A. L., McLaughlin, T., and O'Leary, D. D. (2001). Topographic-specific axon branching controlled by ephrin-As is the critical event in retinotectal map development. *J. Neurosci.* **21,** 8548–8563.

Yates, P. A., Holub, A. D., McLaughlin, T., Sejnowski, T. J., and O'Leary, D. D. M. (2004). Computational modeling of retinotopic map development to define contributions of EphA–ephrinA gradients, axon–axon interactions, and patterned activity. *J. Neurobiol.* **59,** 95–113.

Yoon, M. G. (1976). Progress of topographic regulation of the visual projection in the halved optic tectum of adult goldfish. *J. Physiol.* **257,** 621–643.

Yoon, M. G. (1980). Retention of topographic addresses by reciprocally translocated tectal re-implants in adult goldfish. *J. Physiol.* **308,** 197–215.

Young, J. M., Waleszczyk, W. J., Wang, C., Calford, M. B., Dreher, B., and Obermayer, K. (2007). Cortical reorganization consistent with spike timing-but not correlation-dependent plasticity. *Nat. Neurosci.* **10,** 887–895.

Yu, H., Farley, B. J., Jin, D. Z., and Sur, M. (2005). The coordinated mapping of visual space and response features in visual cortex. *Neuron* **47,** 267–280.

CHAPTER TWO

Synapse Formation in Developing Neural Circuits

Daniel A. Colón-Ramos

Contents

Abstract

The nervous system consists of hundreds of billions of neurons interconnected into the functional neural networks that underlie behaviors. The capacity of a neuron to innervate and function within a network is mediated via specialized cell junctions known as synapses. Synapses are macromolecular structures that regulate intercellular communication in the nervous system, and are the main gatekeepers of information flow within neural networks. Where and when synapses form determines the connectivity and functionality of neural networks. Therefore, our knowledge of how synapse formation is regulated is critical to our understanding of the nervous system and how it goes awry in neurological disorders.

Synapse formation involves pairing of the pre- and postsynaptic partners at a specific neurospatial coordinate. The specificity of synapse formation requires the precise execution of multiple developmental events, including cell fate specification, cell migration, axon guidance, dendritic growth, synaptic target

Program in Cellular Neuroscience, Neurodegeneration and Repair, Department of Cell Biology, Yale University School of Medicine, New Haven, Connecticut, USA

Current Topics in Developmental Biology, Volume 87
ISSN 0070-2153, DOI: 10.1016/S0070-2153(09)01202-2

selection, and synaptogenesis (Juttner and Rathjen in *Cell. Mol. Life Sci.* 62:2811, 2005; Salie *et al.*, in *Neuron* 45:189, 2005; Waites *et al.*, in *Annu. Rev. Neurosci.* 28:251, 2005). Remarkably, during the development of the vertebrate nervous system, these developmental processes occur almost simultaneously in billions of neurons, resulting in the formation of trillions of synapses. How this remarkable specificity is orchestrated during development is one of the outstanding questions in the field of neurobiology, and the focus of discussion of this chapter.

We center the discussion of this chapter on the early developmental events that orchestrate the process of synaptogenesis prior to activity-dependent mechanisms. We have therefore limited the discussion of important activity-dependent synaptogenic events, which are discussed in other chapters of this book. Moreover, our discussion is biased toward lessons we have learned from invertebrate systems, in particular from *C. elegans* and *Drosophila*. We did so to complement the discussions from other chapters in this book, which focus on the important findings that have recently emerged from the vertebrate literature.

The chapter begins with a brief history of the field of synaptic biology. This serves as a backdrop to introduce some of the historically outstanding questions of synaptic development that have eluded us during the past century, and which are the focus of this review. We then discuss some general features of synaptic structure as it relates to its function. In particular, we will highlight evolutionarily conserved traits shared by all synaptic structures, and how these features have helped optimize these ancient cellular junctions for interneural communication.

We then discuss the regulatory signals that orchestrate the precise assembly of these conserved macromolecular structures. This discussion will be framed in the context of the neurodevelopmental process. Specifically, much of our discussion will focus on how the seemingly disparate developmental processes are intimately linked at a molecular level, and how this relationship might be crucial in the developmental orchestration of circuit assembly. We hope that the discussion of the multifunctional cues that direct circuit development provides a conceptual framework into understanding how, with a limited set of signaling molecules, precise neural wiring can be coordinated between synaptic partners.

1. Introduction

1.1. A historical perspective

The history of synapse biology starts at the end of the eighteenth century, with the studies of Luigi Galvani and his descriptions of "animal electricity." In these classical studies, Galvani observed that he could induce the contraction of limb muscles when he inserted a metal hook into the medulla of the frog and attached the other end to an iron railing. These observations marked the first experimental demonstrations of synaptic transmission (Cowan and Kandel, 2001).

Most of the subsequent synaptic studies in the nineteenth century and earlier half of the twentieth century also focused around the functionality of synapses, or synaptic transmission. It is therefore befitting that the actual term "synapse" was not coined by a neuroanatomist, but by a physiologist named Charles Sherrington. Sherrington coined the term "synapse" to refer to the special connections from one nerve cell to another that facilitated the transmission of nervous impulses (Cowan and Kandel, 2001).

While physiologists and neuropharmacologists were functionally defining the concept of synapses, neuroanatomists tangled in a bitter debate on their existence. The main reason for this debate was that during the nineteenth century and earlier part of the twentieth century, nobody could visualize cell membranes and establish conclusively the existence of synapses. However, in spite of these technological limitations, some insightful neurobiologists garnered enough experimental evidence to propose the anatomical existence of synapses.

Most of these early observations came from a specialized synapse: the neuromuscular junction (NMJ). Because of its size, morphology and functional readouts, NMJs informed then, as they do now, most of our knowledge on synaptic biology. Taking advantage of this system, physiologist Willy Kühne and anatomist Wilhelm Krause independently hypothesized the existence of synapses at the site of contact between nerve cells and muscles (Cowan and Kandel, 2001).

The question of the existence of interneuronal synapses was much harder to settle. Synapses in the central nervous system are much smaller than NMJs, in closed apposition to one another and packed at very high densities. This made their visualization with the methods used during nineteenth century downright impossible and triggered the postulation of the "reticular theory": the idea that the nervous system lacked functional separation of nerve cells and was syncytial, rather than synaptic, in nature (Cowan and Kandel, 2001; Westfall, 1996).

The theory turned out, of course, to be wrong. Although this was not conclusively shown until the advent of electron microscopy in the 1950s, the first evidence that neurons were discrete units came from developmental, pathological, and anatomical observations in the nineteenth century. Most notable among these early studies are Santiago Ramón y Cajal's. By using a method derived by Golgi, which stains only 1% of the cells, Cajal was able to visualize the morphology of individual cells in the context of the nervous system. His detailed characterization of neurons not only provided critical evidence for the neuron doctrine, but also stated the "Principios de la Especificidad de la Conexión": the idea that nerve cells connect to each other in a specific fashion to form precise networks (Cowan and Kandel, 2001).

Although it would take another half a century for cell biologists to visualize synapses, Cajal's observations and insights at the turn of the nineteenth century provided the conceptual basis that has driven most of the neurodevelopmental questions since then. Over a century after Cajal's initial descriptions, we are still untangling the complex morass that is the central nervous system and tackling the questions staged by his landmark observations: How are the numerous cell types in the nervous system specified? What directs neurites to connect to each other? What are the cellular and molecular factors that underlie the "Principles of connection specificity"?

1.2. Synaptic structure and function

During the last century however, and thanks in great part to technical advances in cell biology, the field has made great progress in its understanding of the synaptic structure as it relates to synaptic function. Most notably, electron microscopy allowed the visualization of synapses for the first time in the 1950s. This work, spearheaded by George Palade and Keith Porter, provided unequivocal evidence for the neuron doctrine and the existence of synapses, and identified the different types of synapses and their structural components (Cowan and Kandel, 2001; De Camilli *et al.*, 2001).

There are two general categories of synapses: electrical synapses and chemical synapses. Physiologists and neuropharmacologists functionally defined these two categories of synapses well before they were visualized by cell biologists (Cowan and Kandel, 2001). But the cell biological work that proceeded from the physiological studies demonstrated that these two functional categories corresponded to completely different structures. Electrical synapses are gap junctions that allow bidirectional propagation of signals, including electrical stimuli. They allow the fastest mode of electrical propagation across cells, and are now known to be important in synchronizing neural activity across networks (De Camilli *et al.*, 2001). These gap junctions will not be further discussed in this chapter.

Chemical synapses allow communication between discontinuous neurons via the highly regulated secretion of chemical intermediate signals. Unlike electrical synapses, chemical synapses are polarized junctions that allow the flow of information in just a single direction. Because of their highly regulated and directional transfer of information, chemical synapses have been the focus of most of the synaptic biology studies, and as such will remain the focus of our chapter.

Although there is great morphological and molecular variability among chemical synapses, all chemical synapses share common structural and functional features (De Camilli *et al.*, 2001). They consist of two asymmetrically juxtaposed components linking two separate cells: a presynaptic specialization and a postsynaptic region. The presynaptic specializations are specialized regions in the presynaptic cell with an abundance of neurotransmitter-filled synaptic vesicles. Presynaptic specializations also contain the active zone structures that facilitate vesicle fusion and the release of neurotransmitter content to the intersynaptic space, called the synaptic cleft. The postsynaptic region is an area of the postsynaptic cell with a high concentration of neurotransmitter receptors, channels, and downstream signaling molecules. The neurotransmitters released by the presynaptic specializations are sensed by the receptors at the postsynaptic site, activating downstream signaling molecules, opening channels, and propagating the nervous impulse to the postsynaptic partner. These general features of the presynaptic and postsynaptic specializations are shared by all classes of synaptic structures.

The synaptic structure as described above is also very well conserved across evolution. Sea anemones and hydra (Phylum Cnidaria) have the most primitive nervous system, which consists of a diffuse network of neurons. These nerve nets, however, are connected via chemical and electrical synapses that are fully capable of transmitting and regulating information flow (Anderson and Spencer, 1989; Peteya, 1973; Westfall, 1996). Close inspection of these synaptic structures reveal that Cnidarian synapses have similar structural components as those of higher organisms, with defined presynaptic and postsynaptic specializations in close juxtaposition (Anderson and Spencer, 1989; Peteya, 1973; Westfall, 1996).

The presence of a conserved synaptic structure in these primitive nervous systems reveals that synapses are as ancient as the nervous system itself. This evolutionary conservation of the synaptic structure also underscores the importance of these specialized cell junctions in interneuronal communication and the functioning of the neural networks (Anderson and Spencer, 1989; Peteya, 1973; Westfall, 1996).

Interestingly, a recent study suggests that the evolution of the synaptic molecular machine might even precede the evolution of the nervous system (Sakarya *et al.*, 2007). Although sponges (Phylum Porifera) are the only metazoans without a nervous system, it was found that sponges express a nearly complete set of postsynaptic protein homologues that are hypothesized to assemble into synaptic-like scaffolds. Although sponges do not have neurons, these postsynaptic-like structures are hypothesized to act as chemosensory structures capable of responding to environmental cues (Sakarya *et al.*, 2007).

Other molecular components of the presynaptic machine, such as the synaptic vesicle cycle regulators, also predate the existence of the nervous system and are very well conserved across evolution (Sudhof, 2004). It is provocative that these macromolecular machines, presumable "building blocks" of the synapse, might be found even in the absence of a nervous system itself, an observation that underscores the importance and conservation of these signaling complexes throughout evolution (Sakarya *et al.*, 2007).

Molecular and genetic studies in model invertebrate and vertebrate animals have also supported the notion that the ultrastructural conservation of synapses corresponds to a conservation at the molecular level. For instance, in the simple nervous system of the nematode *Caenorhabditis elegans*, which consists of only 302 neurons, the number of neurotransmitters and receptors required for the proper functioning of its ~5000 synapses approaches in complexity those used by the hundreds of trillions of synapses in the vertebrate nervous system (Rand and Nonet, 1997).

This suggests that throughout evolution, the increased capacity of information processing and storage observed in higher organisms is not the result of marked changes in the complexity of the synaptic structure. We speculate that this complexity results from an increasingly sophisticated neural framework in way of the abundance and organization of neural networks. Where, when, and how synapses form during development play critical roles on the wiring and function of neural networks. Although the neural network organization varies vastly across animals, the biological basis of synapses is shared from the simplest networks of Cnidaria to the complex neuropils of the human brain.

2. Synaptogenesis During Development

The organization of where, when, and how synapses are formed plays an instrumental role in directing the connectivity of circuits and organizing the neuroarchitecture that enables information processing, storage, and ultimately behaviors. As such, the developmental questions postulated by Cajal in his "Principles of connection specificity" are of great importance to our understanding of the assembly and function of the nervous system. What are the molecular and cellular factors that direct the precise innervation of hundreds of trillions of synapses during development?

Neural circuit formation requires the intricate orchestration of multiple developmental events including cell fate specification, cell migration, axon guidance, dendritic growth, synaptic target selection, and synaptogenesis (Juttner and Rathjen, 2005; Salie *et al.*, 2005; Waites *et al.*, 2005). The correct innervation of a given circuit requires the successful completion of all of these developmental steps in both synaptic partners. As such, synaptogenesis marks the final step of a complicated developmental dance where, after successful completion of the aforementioned steps, both synaptic partners converge at a specific location to form a specialized junction.

Although the field has now identified a number of molecules required for each of these developmental steps, we know much less about how these different developmental steps act in concert to direct the development of circuits. In the next sections, we will discuss the process of synaptic formation in the context of the complex developmental dance that brings neurons together.

2.1. Neuronal cell fate and synaptogenesis

Circuit formation begins with cell fate specification. During cell fate specification, the seemingly homogeneous neuroepithelium of the developing embryo differentiates into a hugely diverse number of neurons, each tailored morphologically and structurally for its particular functional role. Each of these neuron types has a distinct morphology, axonal and dendritic trajectory, and, of particular interest to this chapter, synaptic property in the way of synaptic specificity and neurotransmitter content. How each neuron adopts a particular identity and how this identity directs its connectivity remain outstanding problems in neurobiology.

Embryological studies have shown that morphogens and transcription factors play crucial roles in the specification of cell fate during the development of neural circuits. The expression of morphogens by discrete tissues establishes gradients along multiple developmental axes. These morphogenic gradients create a unique grid that conveys positional information in the developing embryo. Neural precursor cells respond to this positional information by expressing a specific set of transcription factors. The combinatorial expression of these transcription factors confers the neural precursor cells with an identity that can then be inherited by its descendants (O'Leary *et al.*, 2007). Thus, the extrinsic positional information delineated by a grid of morphogenic gradients is translated into an intrinsic and inheritable cellular identity via the expression of a combinatorial code of transcription factors.

A growing body of literature supports the notion that this combinatorial code of transcription factors can confer important connectivity information to certain neurons (Polleux *et al.*, 2007). For instance, transcription factors have now been shown to be important for proper projections of retinoganglion cells from the retina to the thalamus, for projections of axons from thalamic nuclei to cortical areas and for the patterning of cortical efferent projections (Polleux *et al.*, 2007). But are transcription factors directly required for synaptic specificity? Could the combinatorial transcriptional code also direct circuit innervation at the level of synapse formation?

Several lines of evidence suggest that transcription factors can direct where and how synapses are formed even after the process of axonal and dendritic guidance has concluded. The strongest evidence for the importance of transcription factors directly controlling synaptic specification comes from studies in the motor neurons of the ventral nerve cord of *C. elegans* (Von Stetina *et al.*, 2006). Synapses in *C. elegans* are formed *en passant*, or along the length of the axon. This biological trait allows for an easier developmental dissection of the axon guidance versus the synaptogenesis steps as, unlike end-button synapses, *en passant* synapses are formed on the side of axons, which contact many potential synaptic targets (White *et al.*, 1986). Early genetic studies in the motor neurons of the ventral nerve cord showed that mutant animals lacking the gene *unc-4*, which encodes a Prd-like homeodomain transcription factor, display a strong motor movement defect in backward locomotion. Ultrastructural studies on the innervation and morphology of nerve cord neurons showed that absence of the transcription factor did not alter the organization of the nerve cord, and all neurons looked normal in terms of morphology, position in the nerve cord fascicle and guidance. Interestingly, *unc-4* mutant animals displayed abnormal synaptic specificity, with motor neurons innervating their incorrect partners. This suggested that the Prd-like homeodomain transcription factor UNC-4 directly controls synaptic choice without affecting other neural traits such as outgrowth and fasciculation (Miller *et al.*, 1992; White *et al.*, 1992).

Further studies in other neurons determined that the UNC-4 transcription factor also controls the expression of molecules involved in synaptic strength (Lickteig *et al.*, 2001; Von Stetina *et al.*, 2007). This capacity to regulate the expression of molecules involved in synaptic strength is independent from its capacity to regulate synaptic specificity (Lickteig *et al.*, 2001). Although the identity of the targets of UNC-4 remains unknown, these studies show that transcription factors can directly regulate different aspects of synaptic biology, from formation of the synapse during development to the strength of synaptic connections (Von Stetina *et al.*, 2006).

Another example of how transcription factors can regulate synapse formation is found in the cockroach cercal system. The cerci are an appendages in the rear of the cockroach where filiform hairs are innervated by a single sensory neuron. In newly hatched (first-instar) roaches there are just two sensory neurons that respond to stimuli: the lateral sensory neuron, which responds to stimuli form the front of the animal, and the medial sensory neuron, which responds to stimuli from the rear of the animal (Fig. 2.1). Although the arborization of these two sensory neurons overlaps, the two sensory neurons connect specifically and exclusively to different subsets of giant interneurons. This synaptic specificity directs directional sensitivity to stimuli, allowing the animal to discern if the stimuli come from the front or the rear, and eliciting the corresponding escape response (Blagburn and Bacon, 2004).

Since these two filiform hair afferents have overlapping arborizations, the axonal projections cannot be the primary determinant of synaptic specificity. Instead, synaptic specificity must be directed by additional cues that allow these overlapping arbors to innervate specific targets. Loss-of-function studies in this system showed that the homeodomain transcription

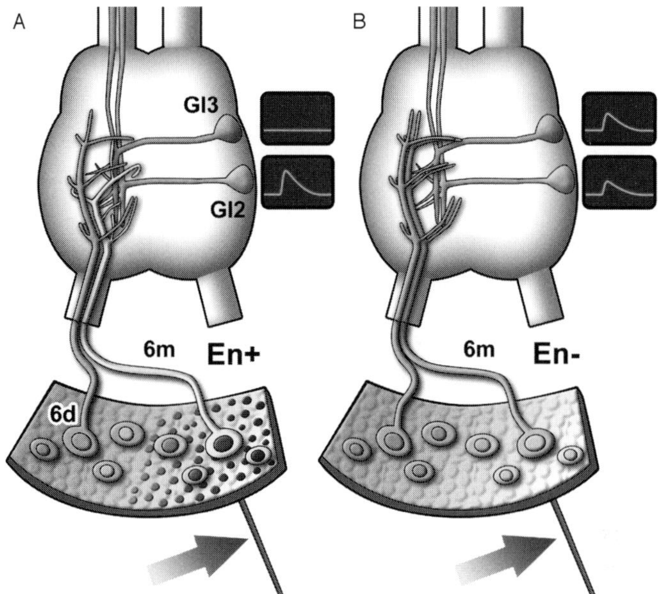

Figure 2.1 Transcription factor *Engrailed* directs synaptic specificity in the cockroach cercal system. Wiring diagram of the cockroach cercal system in the second instar cockroach. (A) In wild-type animals, medial sensory neuron (neuron 6m, in yellow) expresses Engrailed (represented by dark nuclei), while lateral sensory neuron (6d, in blue) does not (represented by clear nuclei). Although the axonal arbors of the medial and lateral sensory neurons overlap, they display specificity by connecting with specific target interneurons: medial sensory neuron connects to interneuron Gl2 (in orange), while lateral sensory neuron connects to interneuron Gl3 (in green). This specificity can be physiologically recorded, so that, for instance, stimuli in the filiform hair linked to the medial sensory neuron (dark arrow) results in depolarization of interneuron Gl2 and not Gl3 (represented by physiological recording to the right of the schematic). (B) Loss of Engrailed by dsRNA disrupts the medial sensory neuron identity and connectivity. In the absence of Engrailed, the medial sensory neuron (6m) adopts an identity similar to the lateral sensory neuron in terms of the branching structure (compare branching schematic of 6m in A (yellow) with 6m in B (blue)). Loss of Engrailed also disrupts medial sensory neuron synaptic specificity (represented by physiological recording to the right of the schematic). (See Color Insert.)

factor Engrailed is critical in directing this specification of synaptic connections (Marie *et al.*, 2000).

Engrailed is expressed by the medial half of the central epidermis of developing animals, including the medial sensory neuron which responds to stimuli from the rear of the animal. Engrailed is not expressed by the lateral sensory neuron that responds to stimuli from the front of the animal. In the absence of Engrailed, the medial sensory neuron adopts a pattern of synaptic connections similar to that of the Engrailed-negative, lateral sensory neuron (Fig. 2.1). These results indicate that transcription factor Engrailed is required for correct specification of synaptic connections (Marie *et al.*, 2000).

Interestingly, persistent expression of Engrailed was also shown to be required for the specification of other developmental traits of the medial sensory neuron. By manipulating Engrailed levels at different developmental stages, the authors went on to show that Engrailed is required in postmitotic neurons to control axon arborization and synaptic specification. They showed that these two events are developmentally separable, but are both dependent on the same transcription factor. These findings demonstrate that Engrailed can direct discrete connectivity decisions at different developmental stages (Marie and Blagburn, 2003; Marie *et al.*, 2002). Furthermore, it highlights the role of transcription factors in coordinating and integrating the different developmental decisions that need to be made to direct neural connectivity.

Therefore, transcription factors can act as conveying points, receiving inputs, and directing different developmental steps that range from cell identity and neurite guidance to synapse assembly. When integrated (as in the case of Engrailed), these different activities could orchestrate the interdependent development and innervation of circuits. Although we understand the importance of transcription factors in directing discrete developmental steps, our knowledge on how their combinatorial and interdependent activity leads to correct innervation is still limited. For instance, in most of the experimental systems in which transcriptional regulation has been shown to affect circuit formation we do not yet know the identity of the guidance or synaptic specificity cues, how these cues are controlled and how their activity directs connectivity.

2.2. Axon guidance and synaptogenesis

Once neural cell fate is specified and neuron precursors have migrated to the appropriate regions, they extend polarized projections that become their axons and dendrites. The axonal processes can extend long distances, navigating complex cellular environments before reaching their postsynaptic partner. This guidance is mediated through the growth cone, a specialized sensing device at the tip of the outgrowing axon. Growth cones express a series of guidance receptors that are capable of sensing a variety of long-range (diffusible) and short-range (surface-bound) guidance cues. These guidance cues, which can be attractive or repulsive, are secreted by guidepost cells and intermediate targets. The spatial and temporal presence of the guidance cues, combined with the expression of the receptors in the growth cone, enables the axon to navigate through the labyrinth that is the developing nervous system to reach its target (Plachez and Richards, 2005; Tessier-Lavigne and Goodman, 1996). Upon reaching and contacting its target, the axon transforms into a presynaptic specialization capable of transducing synaptic signals to the postsynaptic target.

One of the outstanding questions in the field of synaptogenesis is how this transformation is mediated. How does the axon identify its correct postsynaptic target? During guidance, how does the growth cone differentiate between intermediate guidepost targets and its final target? Upon reaching the target region, how does it discriminate between potential partners to innervate its correct postsynaptic partner? And what are the cell biological changes that occur in the axon to transform it into a specialized presynaptic junction?

Clues on how this transformation occurs have come from studies in one of the better-understood systems of growth cone guidance and synaptic targeting: the RP3 motoneuron in the *Drosophila* embryos. The RP3 motoneuron can be visualized during development of the *Drosophila* embryonic CNS with single-cell resolution and in the context of the intact nervous system. This is done by using immunocytochemistry techniques that allow detailed characterization of the developmental decisions made by this motoneuron. These studies demonstrated that the RP3 axon undergoes a stereotypical sequence of guidance events before reaching its final targets, two muscles known as muscles 6 and 7 (Fig. 2.2). Remarkably, upon reaching the target region RP3 comes within filopodial reach of over a dozen different muscles, yet specifically innervates only its correct targets. In wild-type embryos this stereotyped sequence of developmental events, and the innervation of the correct targets, happens with 100% accuracy (Chiba and Rose, 1998).

Genetic studies in the RP3 system showed that the long-range guidance decisions and the short-range synaptic targeting choices are directed by different molecular cues. The RP3 targets, muscles 6 and 7, express the axon guidance cue Netrin. However, deletion of the Netrin gene does not affect long-range guidance decisions of the RP3 axon: in the absence of Netrin or its receptor (Frazzled), the RP3 growth cone makes its normal guidance decisions, exiting the CNS, leaving the nerve bundles and entering the appropriate ventral muscle domain in a timely fashion to reach the neighborhoods of its targets, muscles 6 and 7 (Fig. 2.2). However, although the RP3 growth cone comes within filopodial reach of its targets, in these mutants it fails to innervate them robustly (Mitchell *et al.*, 1996). These studies suggest that Netrin is not required for the long-range guidance decisions of the RP3 motoneuron, but instead is required for its short-range synaptic targeting. Furthermore, these studies indicate that the long-range guidance decisions and the short-range, synaptic targeting decisions of the RP3 motoneuron are two different, genetically separable events that are not dependent on the same molecular factors.

Although the RP3 guidance and synaptic targeting events are genetically separable, short-range targeting in this system is still mediated by a guidance

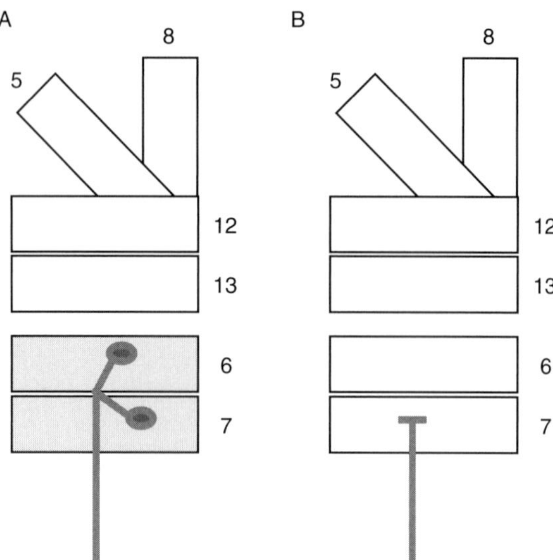

Figure 2.2 Axon guidance molecule Netrin is required for synaptic targeting events in the *Drosophila* embryo. Schematic diagram of RP3 neuron (blue) and the body wall muscles in the *Drosophila* embryo (represented here are muscles 6, 7, 13, 12, 5, and 8). (A) In wild-type animals, muscles 6 and 7 express Netrin (in pink). Muscles 12, 13, 5, and 8 express repulsive cues such as semaphorins, and do not express Netrin (lack of Netrin expression represented in white). Expression of Netrin by muscles 6 and 7 induces short-range targeting and specific innervation of these muscles by the RP3 neuron. (B) In Netrin loss-of-function mutants, the RP3 neuron reaches the neighborhood of muscles 6 and 7 in a timely fashion, coming within filopodial reach of its targets, but fails to innervate these muscles correctly. These studies suggest that Netrin is not required for long-range guidance of the RP3 neuron, but is instead required for its short-range synaptic targeting. (See Color Insert.)

molecule. The Netrin pathway has been shown to regulate cell migration, neural polarization leading to growth cone formation, and axon guidance events (Adler *et al.*, 2006; Kennedy and Tessier-Lavigne, 1995). The molecular nature of Netrin suggests, *prima facie*, that the synaptic targeting events could just be short-range guidance events that refine growth cone steering. However, the experimental data are also consistent with the possibility that Netrin plays a role in transforming the growth cone into a presynaptic structure.

Given Netrin's requirement on directing short-range targeting, it is not possible to conclusively determine if Netrin plays any additional downstream synaptogenic roles in the RP7 motoneuron. The reason for this is that the RP7 motoneuron synapses are terminal button synapses. In terminal button synapses, the final process of synaptic targeting is intimately and seamlessly linked to the synaptic formation step. It is therefore very difficult to separate these two developmental events.

Nonetheless, another class of synapses, *en passant* synapses, provides a way of developmentally differentiating between short-range guidance events, such as synaptic targeting, and bona fide synapse formation. Unlike terminal button synapses, *en passant* synapses are formed along the length of the axon. Because it is the axonal shaft, and not the growth cone, which transforms into a presynaptic specialization, *en passant* synapses do not require the short-range guidance events that target the growth cones prior to synaptic assembly. Instead, during *en passant* synapse formation, guidance and fasciculation events bring the axonal shaft in contact with the potential postsynaptic targets. Once this neural framework is established, downstream synaptic specification events direct the formation of synaptic structures along the axonal shaft, discriminating between the fasciculating neurites to innervate the correct synaptic partners. For *en passant* synapse formation, contact between fasciculating neurites is necessary, but not sufficient to direct synaptic formation, indicating the existence of downstream synaptic specificity events that direct cell–cell recognition and synapse formation (White *et al.*, 1986). The cytoarchitecture of *en passant* synapses facilitates the identification, and separation, of events involved in synaptic specificity from those involved in axon targeting.

By studying the *en passant* synapses in the nematode *C. elegans*, the Netrin pathway was found to play a role in directing the formation of presynaptic specializations different from its conventional guidance role. This was done by visualizing the developmental decisions that led to the innervation of two interneurons (AIY and RIA) in the nematode nerve ring. Similar to the RP3 system, this system allowed visualization of the developmental decisions of these neurons with single-cell resolution and in the context of the intact nervous system, but also *in vivo* (Gray *et al.*, 2005; Mori and Ohshima, 1995). It was observed that AIY must undergo a series of stereotyped guidance decisions to reach and contact RIA. AIY and RIA then connect to each other through *en passant* synapses formed at discrete regions of their respective processes (White *et al.*, 1986). The resulting AIY: RIA circuit assembles in a stereotyped fashion in 100% of wild-type animals (Fig. 2.3).

In the absence of Netrin or the Netrin receptor, the majority of AIY cells still make their guidance decisions correctly. In spite of the fact that the guidance decisions are normal in the majority of animals, AIY fails to properly form presynaptic specializations at the usual site of contact with RIA (Fig. 2.3). Cell-autonomous rescue of the Netrin receptor in the presynaptic neuron AIY is sufficient to rescue the presynaptic patterning defect, and when the subcellular localization of the Netrin receptor was visualized, it was observed that the Netrin receptor was enriched at the presynaptic sites. Together, these results indicate that the Netrin pathway plays a role in organizing synaptogenesis and that this role is independent of its function in guidance (Colon-Ramos *et al.*, 2007).

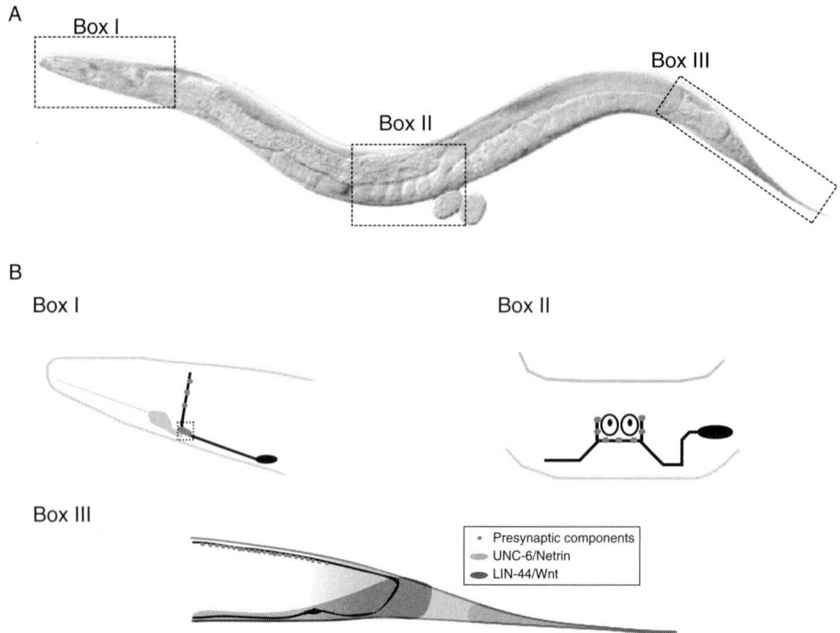

Figure 2.3 Guidepost cells direct synaptic specification. Schematic representing examples of guidepost cells directing synaptic specification in *C. elegans*. (A) Image of *C. elegans* with the discussed regions boxed. (B) *Box-I*: In the nematode head, interneuron AIY (black) contacts many neurons, but connects specifically to interneuron RIA (not shown) at a subcellular region of its neurite (boxed). This specificity is directed by ventral cephalic sheath cells (pink). Ventral cephalic sheath cells are glial cells that project a process posteriorly, where it contacts AIY and RIA, and also express Netrin. The expression of Netrin directs presynaptic assembly (in green) in the correct subcellular region of AIY (boxed region). *Box-II*: In the nematode egg-laying circuit, neuron HSNL (in black) innervates other neurons and muscles (not shown) in a specific and stereotyped fashion. The postsynaptic partners of HSNL are not required for the precise assembly of presynaptic specializations (in green) in a subcellular region of HSNL. Instead, guidepost epithelial cells (dark spheres) express an immunoglobulin superfamily receptor (SYG-2) that directs where synapses form in HSNL. *Box-III*: In the posterior part of the nematode, neuron DA9 (in black) elaborates a dendrite anteriorly within the ventral nerve cord and extends an axon commissurally and then longitudinally along the dorsal nerve cord. UNC-6/Netrin (pink) and LIN-44/Wnt (blue) direct synaptic specification (green) by inhibiting formation of synapses from discrete subcellular domains. Expression of *lin-44/wnt* (blue) by cells in the posterior part of the animal prevents ectopic synapse formation in the commissure, while expression of Netrin (pink) by ventral cells excludes presynaptic components from the ventral dendrite. (See Color Insert.)

A recent report on the identification of a novel interactor in the Netrin pathway supports this newfound role of Netrin in organizing presynaptic assembly. CLEC-38, a transmembrane protein with C-type-lectin-like domains, was recently shown to both regulate outgrowth activity of growth cones, and the organization of presynaptic terminals. In some developmental contexts CLEC-38 acts by inhibiting the Netrin receptor, thereby regulating neural outgrowth during guidance. In other developmental contexts, however, CLEC-38 does not regulate outgrowth, but is instead required for the organization of presynaptic terminals. Although it is not yet known if this presynaptic role of CLEC-38 is also mediated via the Netrin receptor, these data showed that CLEC-38, a regulator of Netrin activity and axon outgrowth, plays additional roles in organizing presynaptic specializations (Kulkarni *et al.*, 2008).

Other families of guidance molecules have also been observed to play synaptogenic roles depending on the developmental context. For instance, in vertebrates, guidance molecules such as the Eph family of receptors and their ephrin ligands have been shown to play roles in growth cone guidance as well as the development of mature excitatory synapses (Dalva *et al.*, 2000). The distinct cellular responses of these different developmental events are also likely generated by the developmental context and by diverse downstream targets (Murai and Pasquale, 2004). Another family of receptors, the LAR-like phosphatase receptors, has also been shown to function at the level of axon guidance and presynaptic organization. Cell biological and genetic characterizations of this receptor showed that in nematodes, these two different activities are regulated by differentially spliced isoforms of the same receptor (Ackley *et al.*, 2005).

The existence of these shared pathways underscores the molecular link between guidance and synaptogenesis. It also provides a conceptual framework on how growth cones, upon reaching their synaptic targets, could transition from being an outgrowth structure to a presynaptic terminal. But how can the same protein "molecularly multitask" and direct different developmental functions? How can the same receptor and ligand elicit diverse cellular responses?

In the case of the Netrin receptor, for instance, the same molecule can regulate cell migration, neuronal polarization, axon guidance, and synapse formation in different developmental contexts. Although these events have very different outcomes, there is an underlying similarity at the cell biological level: all of these events impinge on a polarization process that restructures the cytoskeleton. For example, during the maturation of the *C. elegans* neuron HSN, Netrin restricts the Netrin receptor localization to the neuronal side facing the source of Netrin. This leads to the polarized formation of the HSN growth cone (Adler *et al.*, 2006). During guidance, Netrin polarizes the growth cone cytoskeleton to generate directed growth (Kennedy and Tessier-Lavigne, 1995). Similarly, in the AIY interneuron Netrin induces a localized polarization event to transform a region of the axon shaft into a specialized presynaptic area (Colon-Ramos *et al.*, 2007).

It is not well understood how downstream factors and regulators parcel out these signals to result in different developmental outcomes. However, with our increasing molecular understanding of these developmental events, it has become clear that guidance and synaptogenesis are intimately linked at a molecular level. Future work on these signal transduction pathways will allow us to understand how the same ligand/receptor pair can elicit distinct developmental outcomes, and how these events can be orchestrated in different cells to enable circuit assembly.

3. Building a Synapse

The process of guidance and target recognition is followed by synapse formation. How synapses are assembled is a formidable developmental question in its own right. As discussed previously, synapses need to form onto the right partner, at the right density, and at a specific subcellular location with respect to the dendrites. Moreover, the assembly of presynaptic sites also needs to match the postsynaptic densities in terms of localization and identity of the neurotransmitter and postsynaptic receptor (Juttner and Rathjen, 2005).

3.1. Cell adhesion in synaptic assembly

Synaptogenesis can be subdivided into two developmentally distinct steps (1) synaptic specificity and (2) synaptic assembly. Synaptic specificity describes the process that directs *where* synapses form: from the selection of the right partner to the formation of synapses at the right subcellular compartment. Synaptic assembly describes *how* synapses are formed: from the assembly of the macromolecular presynaptic structure to the formation of the signaling-rich postsynaptic specializations.

How are these two processes integrated to result in correct synaptic development? The genesis of the synapse officially starts with the contact and communication between the pre- and the postsynaptic partners. Therefore, synaptic specificity is traditionally thought to be determined by cell surface molecules that mediate this synaptic partner interaction. The classical model states that contact between correct partners, mediated via cell adhesion molecules, would then trigger inductive events that lead to the assembly and/or differentiation of pre- and postsynaptic specializations (Waites *et al.*, 2005).

Spurred by this classic model, a number of studies have focused on the identification of synaptogenic cell surface molecules. These studies led to the identification of cell adhesion molecules that direct a variety of events in synaptic biology. The nature and importance of these cell–cell signaling molecules in synapse biology have been discussed elsewhere (Akins and Biederer, 2006; Benson *et al.*, 2001; Dalva *et al.*, 2007; Juttner and Rathjen, 2005;

Scheiffele, 2003; Yamagata *et al.*, 2003), so in this section we will only provide a very brief summary of the conceptual findings stemming from these studies.

The molecules identified in these studies fall into four functional categories: they either (1) promote stability by linking synaptic partners, (2) direct target recognition, (3) regulate differentiation of pre- and postsynaptic specializations, or (4) modulate synaptic structure and function (Yamagata *et al.*, 2003). For instance, cadherins have been shown to localize to puncta adherentia and direct synaptic morphology (Scheiffele, 2003). Immunoglobulin superfamily (IgSF) adhesion molecules Dscam, DscamL, Sidekick-1, and Sidekick-2 direct lamina-specific connectivity between specific interneurons and retinal ganglion cells in the vertebrate retina (Yamagata *et al.*, 2002). In *Drosophila*, LRR transmembrane protein *capricious* directs target specificity between muscle 12 and the motoneurons that innervate it (Shishido *et al.*, 1998). Ephrins, on the other hand, can act through the EphB receptor to induce the clustering of NMDA receptors and postsynaptic development (Dalva *et al.*, 2000).

Interestingly, despite the focused efforts of identifying cell adhesion molecules directly involved in synaptogenesis, only two adhesion molecules have been shown to induce formation of presynaptic specializations: neuroligins and SynCAM1 (Akins and Biederer, 2006). The limited number of identified cell adhesion molecules capable of directly regulating synapse formation suggests that additional cues remain to be discovered. These findings also beg the question of how the connectivity of hundreds of trillions of synaptic connections are specified with a limited number of molecular cues.

3.2. Assembling the synaptic components

The classical view of synaptogenesis suggests that upon synaptic contact between partners, cell adhesion molecules induce the assembly of pre- and postsynaptic specifications. This places assembly downstream of the adhesion-mediated specification events. Nonetheless, in a number of developmental contexts *in vivo*, synaptic assembly occurs prior to synaptic contact. For instance, during muscle development in vertebrates, AChR clusters concentrate into high density "hotspots" well before the growing axon has arrived. This postsynaptic clustering of AChR receptors is also observed in aneural myotube cultures and in muscles of animals that have been genetically rendered aneural (Kummer *et al.*, 2006). These experiments indicate that postsynaptic AChR clusters can occur prior to synaptic contact and in the absence of presynaptic neural factors.

Presynaptic specializations can also form prior to cell–cell contact between synaptic partners. For instance, detailed ultrastructural studies in *Xenopus laevis* tadpoles revealed that presynaptic specializations develop prior to the association of the axon with the dendrites (Vaughn, 1989).

These observations are also supported by ultrastructural studies in the developing cortex of vertebrates, which revealed the existence of pre- and postsynaptic specializations that formed prior to the contact between synaptic partners (Craig and Lichtman, 2001).

Tissue culture studies have also supported the notion that synaptic contact is not necessary for the establishment of pre- and postsynaptic specializations. For instance, studies in dissociated hippocampal neurons have demonstrated that, prior to cell–cell contact, functional NMDA and non-NMDA-type glutamate receptors are present on the cell surface (Craig and Lichtman, 2001). Functional studies indicate that these "free" pre- and postsynaptic structures contain the core molecular machinery necessary for their function (Krueger *et al.*, 2003).

Together, these studies indicate some important aspects of the process of synapse formation. First, they demonstrate that synaptic partners are not necessary for the assembly of synaptic components: both the release machinery in presynaptic structures and the neurotransmitter receptor clusters in postsynaptic structures can be established independent of one another. Second, these experiments highlight the developmental and the genetic separation between synaptic assembly and synaptic specificity.

This functional separation is further underscored in tissue culture systems that can reconstitute assembly, but not specificity events. In tissue culture systems, neurons are dissociated, plated, and allowed to form synaptic connections onto neighboring neurons. The dissociation of neurons disrupts the architecture of the nervous system, thereby destroying much of the positional information which mediates synaptic specificity. Nonetheless, dissociated neurons still retain the ability to form synapses to neighboring neurons, to themselves, or even onto polylysine-coated beads (Vaughn, 1989). These data suggest that assembly and specificity events are likely mediated through distinct signal transduction pathways.

Given the genetic separation between these events, how are they linked to enable the assembly of a precisely wired nervous system? For instance, is the observed assembly of "half-synapses" a transient and nonspecific feature of neuronal development? Can they actually influence where synapses will be ultimately formed?

A number of studies have now shown that these "half-synapses" can participate in directing where synapses form. Neurodevelopmental studies in zebrafish embryos showed that postsynaptic AChR aggregates formed in advance of growing axons. Although some aggregates dispersed before innervation, surprisingly, filopodia were stabilized when they contacted the AChR aggregates (Kummer *et al.*, 2006). Furthermore, in dissociated hippocampal cultures, preformed postsynaptic scaffold protein complexes, containing PSD-95, GKAP, Shank, and neuroligin 1, localized to predefined postsynaptic hotspots. Upon contact with axons, these scaffolding complexes induced the recruitment of synaptophysin-containing transport vesicles and the formation of presynaptic specializations (Gerrow *et al.*, 2006). These results suggest that assembly prior to synaptic contact can later influence where synapses will form.

3.3. Guidepost cells, morphogens, and connectivity

The studies discussed in the previous section indicate that the assembly of synapses does not require contact between synaptic partners. Moreover, given the role of these preformed specializations in directing where synapses form, these data suggest that in some developmental contexts, the specification of *where* synapses form is also not dependent on the contact between synaptic partners. Which molecular mechanisms direct synaptic specificity in these contexts?

Accumulating evidence suggests that both intrinsic and extrinsic mechanisms can influence where synapses are formed. For instance, studies in dissociated cortical neuronal cultures revealed that initial formation of presynaptic terminals preferentially occurs at predefined sites within the axonal shaft. In these studies, time-lapse imaging was conducted to track the movement of synaptic vesicle protein transport vesicles (STVs), an organelle containing synaptic vesicle-associated proteins which gets recruited to nascent synapses. It was observed that, even in the absence of postsynaptic partners or glia contact, STVs paused at predefined sites. Upon contact with presumptive postsynaptic partners, presynaptic terminals developed specifically at these predefined sites. Moreover, these sites promoted the formation of stable contacts with dendritic filopodia (Sabo *et al.*, 2006). These studies indicate that intrinsic, predefined pause sites in axon shafts can influence the development of nascent synapses at particular sites along the axon in dissociated neurons.

Extrinsic signals generated by guidepost cells can also provide cues that direct where synapses are assembled. For instance, in *C. elegans* the egg-laying motor neuron (HSNL) specifically innervates muscles and VC neurons in the vulva region of the animal (Fig. 2.3). Surprisingly, the postsynaptic partners are not required for correct formation of presynaptic specializations in HSNL. Instead, guidepost epithelial cells provide a positional cue that directs HSNL presynaptic assembly. This is molecularly mediated through a pair of IgSF proteins, SYG-1 and SYG-2. In *syg-1* or *syg-2* mutants, presynaptic neuron HSNL contacts its normal synaptic partners but fails to form synaptic connections with them. Instead, ectopic synapses are formed onto abnormal postsynaptic targets. SYG-1 and SYG-2 both localize to synapses and bind to each other, acting as a receptor and a ligand. SYG-1 functions cell autonomously in the presynaptic neuron, while SYG-2 functions in the guidepost epithelial cells that are essential for the correct formation of HSNL synapses (Shen and Bargmann, 2003; Shen *et al.*, 2004).

Glial cells can also serve as guideposts in directing the innervation of neurons. As discussed earlier, in the *C. elegans* nerve ring, two interneurons (RIA and AIY) reliably innervate each other at stereotyped locations (Fig. 2.3). A pair of nearby glia-like cells serve as guideposts for the innervation of these two interneurons (Colon-Ramos *et al.*, 2007). In the vertebrate cerebellum, glia cells direct the innervation of two interneurons: stellate and Purkinje cells. Stellate cells exclusively innervate the Purkinje neuron dendrites, and this precision is critical for the proper functioning of these cerebellar GABAergic circuits (Fig. 2.4). It was observed that stellate cells associated with Bergmann glia (BG) during development, and followed the glia process by extending their axon through the curving contours of the BG fibers. By following the guidepost BG fibers, stellate cell processes are able to reach their targets: the dendrite of the Purkinje neurons (Ango *et al.*, 2008).

The factor required in both BG and stellate cells for the proper development of this circuit is an L1 family immunoglobulin cell adhesion molecule, CHL1. Interestingly, previous work had shown that another member of the

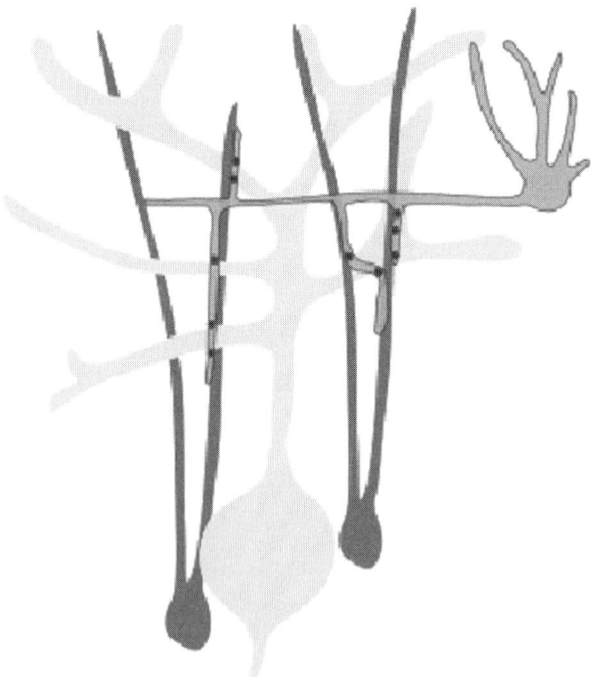

Figure 2.4 Bergmann glia direct the innervation of Stellate axons to the Purkinje dendrites. Purkinje neurons (yellow) are innervated by stellate interneurons (green) exclusively at the dendrites. This precision is directed by Bergmann glia (red), which act as guideposts by directing the stellate interneuron process to their Purkinje neuron targets (synapses in blue). (See Color Insert.)

L1 family, neurofascin186, is required for the specification of another part of this GABAergic circuit: the innervation of the Basket cells and the Purkinje cell axon initial segment (Ango *et al.*, 2004). This molecular characterization of the cerebellar GABAergic circuit suggests that different members of the L1CAM protein family contribute to circuit formation through their cell-specific expression in subsets of neurons and guidepost glial cells.

Some neurons in the vertebrate hippocampus and cortex can also act as guideposts and direct synaptogenesis. For instance, the transient population of Cajal-Retzius cells in the hippocampus serves as a placeholder to facilitate the meeting of the appropriate pre- and postsynaptic cells (Del Rio *et al.*, 1997). Also, during the development of the mammalian cortex, the subplate neurons display a similar guidepost function to arrange the connectivity between the thalamic axons and the layer 4 cortical neurons (McConnell *et al.*, 1989). The significance of these guidepost cells was demonstrated by ablating the guidepost cells and showing a synaptic connectivity defect in ablated animals (Del Rio *et al.*, 1997; Ghosh *et al.*, 1990).

Tissues can also provide inhibitory signals to direct the formation of synapses with subcellular precision. In the DA9 motoneuron of *C. elegans*, Wnt *lin-44* is secreted by four hypodermal cells in the tail. This expression localizes receptor *lin-17*/Frizzled (Fz) to a subdomain of DA9, near the posterior part of the neurite adjacent to the hypodermal cells. This part of the neurite where *lin-17*/Fz localizes is normally devoid of presynaptic specializations (Fig. 2.3). When the Wnt pathway is compromised, however, synapses develop ectopically in this subdomain. Moreover, overexpression of WNT *lin-44* in the posterior part of the animal expands LIN-17 localization and inhibits presynaptic assembly in these new sites of ectopic LIN-17 localization (Klassen and Shen, 2007). The Wnt pathway was also shown to act as a local repressive cue to direct target specificity in *Drosophila*. Studies in embryonic motor neurons showed that Wnt4 acts via Frizzled 2, Derailed-2, and Disheveled to generate target specificity by preventing synapse formation onto nontarget muscles (Inaki *et al.*, 2007).

Wnts can also stimulate the formation of synapses in both vertebrates and invertebrates (Salinas and Zou, 2008). In *Drosophila*, Wnt/wingless is required for the correct development of presynaptic boutons, in terms of both their numbers and their structure, and this activity occurs in a transcription-independent manner (Miech *et al.*, 2008). These observations, together with studies showing a role of Wnts in postsynaptic activation of Frizzled receptors, indicate that Wnt signaling can alter synaptic development by simultaneously modulating the development of presynaptic and postsynaptic structures (Miech *et al.*, 2008).

But how can Wnts both promote and inhibit the formation of synapses? Recent studies in dissociated hippocampal cultures have demonstrated that the opposing effects of Wnts on synapse formation depend on different Wnt ligands. Interestingly, the different ligands differentially activate either the canonical or the noncanonical pathways: activation of the canonical pathway promotes synapse formation, while activation of the noncanonical one inhibits synapse formation (Davis *et al.*, 2008).

Together, these studies demonstrate that in both vertebrates and invertebrates, morphogenic signals such as Wnts can spatially regulate the patterning of synaptic connections. This is achieved by subdividing the neurite into discrete domains and either preventing or promoting synapse formation at specific subcellular compartments.

Other extrinsic cues can also direct neuronal polarity that ultimately impinges on the site of synaptic assembly *in vivo*. In a study that also used the DA9 motoneuron system, it was shown that the axon guidance cue UNC-6/netrin and its receptor UNC-5 act to exclude synaptic vesicles and active zone proteins from the dendrite of DA9 (Fig. 2.3). In *unc-6*/netrin and *unc-5* loss-of-function mutants, presynaptic components mislocalize to the DA9 dendrite, where the level of endogenous UNC-6/netrin is high. In addition, ectopically expressed UNC-6/netrin, acting through UNC-5, was sufficient to exclude endogenous synapses from adjacent subcellular domains within the DA9 axon. Interestingly, this antisynaptogenic activity was interchangeable with that of LIN-44/Wnt (Poon *et al.*, 2008). This suggests that extracellular cues such as netrin and Wnts not only guide axon navigation but also regulate the polarized accumulation of presynaptic components through local exclusion.

Together, these studies indicate that extrinsic signals generated from guidepost cells or neighboring tissues can direct the site of synaptic assembly, thereby modulating synaptic specificity. Although these examples indicate that synaptic contact is not required for the initial specification of synaptogenesis, it should be noted that contact with a postsynaptic-like substrate is required for the eventual stabilization and perseverance of many of these "free" pre- and postsynaptic sites (Vaughn, 1989).

Contact and communication between the synaptic partners is also required for the regulation of the size and shape of the synapse, a process known as synaptic homeostasis (Keshishian, 2002). During synaptic homeostasis, a crosstalk between pre- and postsynaptic specializations takes place across the synaptic cleft. Depending on the developmental context, this crosstalk can be mediated by a number of different molecular cues, which includes fibroblast growth factors (FGFs), bone morphogenetic protein (BMP), signal transduction pathways, and MEF-2 transcriptional responses (Aberle *et al.*, 2002; Goold and Davis, 2007; Salinas, 2005; Simon *et al.*, 2008). The trans-synaptic communication between partners elicits a coordinated regulation of synaptic size, shape, and functionality.

Contact between synaptic partners is also required for the proper modulation of synaptic activity. Although in this chapter we focused our discussion on the early synaptogenic decisions preceding synaptic activity, it should be noted that it is this synaptic activity that, in most developmental contexts, ultimately regulates the stabilization or elimination of many synapses (Flavell and Greenberg, 2008). The early developmental decisions discussed in this chapter generate a neural framework over which activity-dependent changes occur. The activity-dependent regulation of synaptic biology is required, in both the developing and the mature nervous systems, for the maturation, refinement, and plasticity of synaptic connections.

4. Perspective

Correct circuit formation requires an intricate orchestration of multiple developmental events including cell migration, axon guidance, dendritic growth, target selection, and synaptogenesis (Juttner and Rathjen, 2005; Salie *et al.*, 2005; Waites *et al.*, 2005). These events are integrated to enable correct synapse formation between neuronal partners. The developmental innervation of synaptic partners results in hundreds of trillions of precisely wired synaptic connections. Since the human genome has an estimated 25,000 genes, and not all genes are involved in synaptogenesis, these events need to be simultaneously orchestrated in billions of neurons using a limited set of molecular cues.

Genetic, biochemical, and cell biological studies have revealed some of the molecular cues that regulate these developmental processes. Studies of these signaling molecules have revealed classical roles for different protein families. For instance, morphogens, which are known to create gradients with important positional information, are critical for the specification of cell fate decisions. Transmembrane proteins that recognize diffusible factors have been shown to direct processes such as dendritic and axonal outgrowth and guidance. On the other hand, cell–cell adhesion proteins can control synaptogenesis. This has led to the conceptual understanding that gene families with evolutionarily conserved functions could play modular roles in patterning the nervous system (Salie *et al.*, 2005).

Although different protein families can play distinct roles at discrete developmental steps, this modular model of nervous system development does not reflect the complexity of this process *in vivo*. A growing body of literature shows that molecular cues, far from playing exclusive roles at discrete steps, are instead capable of "molecular multitasking" (Salie *et al.*, 2005). For example, morphogenic proteins such as Sonic hedgehog, Wnts, FGF, and BMP have long been known to direct neuronal cell fate by eliciting transcriptional programs. More recent studies have demonstrated that, depending on the cellular context, these canonical morphogens can also provide instructive, transcription-independent signals to control processes such as axon outgrowth, neuronal cell polarity, and synapse formation (Salie *et al.*, 2005). Additionally, signal transduction cascades that have been traditionally thought of as long-range guidance cues have now also been observed to regulate synaptic formation events.

It is provocative to think that "molecular multitasking" could have profound implications for the development of neural circuits. For instance, through the same receptor:ligand pair, multiple signal transduction pathways could be simultaneously activated in different cells with different developmental outcomes. One could speculate that this would allow a single molecular cue to simultaneously direct several independent developmental outcomes in different cells, thereby coordinating circuit assembly by orchestrating the innervation of multiple partners.

The identification of molecular factors and signal transduction cascades involved in synapse formation, combined with an increased understanding of how these molecular factors are integrated to direct circuit formation *in vivo*, will provide us with an increasingly clear picture on how precise synaptogenesis is orchestrated during nervous system development.

ACKNOWLEDGMENTS

I thank M. Hammarlund, S. Margolis, M. Margeta, and G. Maro for thoughtful comments concerning this chapter. I particularly thank K. Shen for helpful discussions, generous advice, and thoughtful comments on the chapter. I also thank J. Blagburn, V. Poon, and F. Ango for contributing images. I apologize to those whose work I did not cite here due to oversight or space constrains. During the preparation of this chapter, I was supported by NIH grant 4R00NS057931-03.

REFERENCES

Aberle, H., Haghighi, A. P., Fetter, R. D., McCabe, B. D., Magalhaes, T. R., and Goodman, C. S. (2002). Wishful thinking encodes a BMP type II receptor that regulates synaptic growth in *Drosophila*. *Neuron* **33,** 545–558.

Ackley, B. D., Harrington, R. J., Hudson, M. L., Williams, L., Kenyon, C. J., Chisholm, A. D., and Jin, Y. (2005). The two isoforms of the *Caenorhabditis elegans* leukocyte-common antigen related receptor tyrosine phosphatase PTP-3 function independently in axon guidance and synapse formation. *J. Neurosci.* **25,** 7517–7528.

Adler, C. E., Fetter, R. D., and Bargmann, C. I. (2006). UNC-6/Netrin induces neuronal asymmetry and defines the site of axon formation. *Nat. Neurosci.* **9,** 511–518.

Akins, M. R., and Biederer, T. (2006). Cell-cell interactions in synaptogenesis. *Curr. Opin. Neurobiol.* **16,** 83–89.

Anderson, P. A., and Spencer, A. N. (1989). The importance of cnidarian synapses for neurobiology. *J. Neurobiol.* **20,** 435–457.

Ango, F., di Cristo, G., Higashiyama, H., Bennett, V., Wu, P., and Huang, Z. J. (2004). Ankyrin-based subcellular gradient of neurofascin, an immunoglobulin family protein, directs GABAergic innervation at purkinje axon initial segment. *Cell* **119,** 257–272.

Ango, F., Wu, C., Van der Want, J. J., Wu, P., Schachner, M., and Huang, Z. J. (2008). Bergmann glia and the recognition molecule CHL1 organize GABAergic axons and direct innervation of Purkinje cell dendrites. *PLoS Biol.* **6,** e103.

Benson, D. L., Colman, D. R., and Huntley, G. W. (2001). Molecules, maps and synapse specificity. *Nat. Rev. Neurosci.* **2,** 899–909.

Blagburn, J. M., and Bacon, J. P. (2004). Control of central synaptic specificity in insect sensory neurons. *Annu. Rev. Neurosci.* **27,** 29–51.

Chiba, A., and Rose, D. (1998). "Painting" the target: How local molecular cues define synaptic relationships. *Bioessays* **20,** 941–948.

Colon-Ramos, D. A., Margeta, M. A., and Shen, K. (2007). Glia promote local synaptogenesis through UNC-6 (netrin) signaling in *C. elegans*. *Science* **318,** 103–106.

Cowan, W. M., and Kandel, E. R. (2001). A brief history of synapses and synaptic transmittion. *In* "Synapses" (T. C. Sudhof, W. M. Cowan, and C. F. Stevens, Eds.), pp. 1–88. The Johns Hopkins University Press, Baltimore.

Craig, A. M., and Lichtman, J. W. (2001). Synapse formation and maturation. *In* "Synapses" (T. C. Sudhof, W. M. Cowan, and C. F. Stevens, Eds.), pp. 1–88. The Johns Hopkins University Press, Baltimore.

Dalva, M. B., McClelland, A. C., and Kayser, M. S. (2007). Cell adhesion molecules: Signalling functions at the synapse. *Nat. Rev. Neurosci.* **8,** 206–220.

Dalva, M. B., Takasu, M. A., Lin, M. Z., Shamah, S. M., Hu, L., Gale, N. W., and Greenberg, M. E. (2000). EphB receptors interact with NMDA receptors and regulate excitatory synapse formation. *Cell* **103,** 945–956.

Davis, E. K., Zou, Y., and Ghosh, A. (2008). Wnts acting through canonical and noncanonical signaling pathways exert opposite effects on hippocampal synapse formation. *Neural Develop.* **3,** 32.

De Camilli, P., Haucke, V., Takei, K., and Mugnaini, E. (2001). The structure of synapses. *In* "Synapses" (T. C. Sudhof, W. M. Cowan, and C. F. Stevens, Eds.), pp. 1–88. The Johns Hopkins University Press, Baltimore.

Del Rio, J. A., Heimrich, B., Borrell, V., Forster, E., Drakew, A., Alcantara, S., Nakajima, K., Miyata, T., Ogawa, M., Mikoshiba, K., Derer, P., Frotsher, M., and Soriano, E. (1997). A role for Cajal-Retzius cells and reelin in the development of hippocampal connections. *Nature* **385,** 70–74.

Flavell, S. W., and Greenberg, M. E. (2008). Signaling mechanisms linking neuronal activity to gene expression and plasticity of the nervous system. *Annu. Rev. Neurosci.* **31,** 563–590.

Gerrow, K., Romorini, S., Nabi, S. M., Colicos, M. A., Sala, C., and El-Husseini, A. (2006). A preformed complex of postsynaptic proteins is involved in excitatory synapse development. *Neuron* **49,** 547–562.

Ghosh, A., Antonini, A., McConnell, S. K., and Shatz, C. J. (1990). Requirement for subplate neurons in the formation of thalamocortical connections. *Nature* **347,** 179–181.

Goold, C. P., and Davis, G. W. (2007). The BMP ligand Gbb gates the expression of synaptic homeostasis independent of synaptic growth control. *Neuron* **56,** 109–123.

Gray, J. M., Hill, J. J., and Bargmann, C. I. (2005). Inaugural article: A circuit for navigation in *Caenorhabditis elegans*. *Proc. Natl. Acad. Sci. USA* **102,** 3184–3191.

Inaki, M., Yoshikawa, S., Thomas, J. B., Aburatani, H., and Nose, A. (2007). Wnt4 is a local repulsive cue that determines synaptic target specificity. *Curr. Biol.* **17,** 1574–1579.

Juttner, R., and Rathjen, F. G. (2005). Molecular analysis of axonal target specificity and synapse formation. *Cell. Mol. Life Sci.* **62,** 2811–2827.

Kennedy, T. E., and Tessier-Lavigne, M. (1995). Guidance and induction of branch formation in developing axons by target-derived diffusible factors. *Curr. Opin. Neurobiol.* **5,** 83–90.

Keshishian, H. (2002). Is synaptic homeostasis just wishful thinking? *Neuron* **33,** 491–492.

Klassen, M. P., and Shen, K. (2007). Wnt signaling positions neuromuscular connectivity by inhibiting synapse formation in *C. elegans*. *Cell* **130,** 704–716.

Krueger, S. R., Kolar, A., and Fitzsimonds, R. M. (2003). The presynaptic release apparatus is functional in the absence of dendritic contact and highly mobile within isolated axons. *Neuron* **40,** 945–957.

Kulkarni, G., Li, H., and Wadsworth, W. G. (2008). CLEC-38, a transmembrane protein with C-type lectin-like domains, negatively regulates UNC-40-mediated axon outgrowth and promotes presynaptic development in *Caenorhabditis elegans*. *J. Neurosci.* **28,** 4541–4550.

Kummer, T. T., Misgeld, T., and Sanes, J. R. (2006). Assembly of the postsynaptic membrane at the neuromuscular junction: Paradigm lost. *Curr. Opin. Neurobiol.* **16,** 74–82.

Lickteig, K. M., Duerr, J. S., Frisby, D. L., Hall, D. H., Rand, J. B., and Miller, D. M. 3rd. (2001). Regulation of neurotransmitter vesicles by the homeodomain protein UNC-4 and its transcriptional corepressor UNC-37/groucho in *Caenorhabditis elegans* cholinergic motor neurons. *J. Neurosci.* **21,** 2001–2014.

Marie, B., Bacon, J. P., and Blagburn, J. M. (2000). Double-stranded RNA interference shows that Engrailed controls the synaptic specificity of identified sensory neurons. *Curr. Biol.* **10,** 289–292.

Marie, B., and Blagburn, J. M. (2003). Differential roles of engrailed paralogs in determining sensory axon guidance and synaptic target recognition. *J. Neurosci.* **23,** 7854–7862.

Marie, B., Cruz-Orengo, L., and Blagburn, J. M. (2002). Persistent engrailed expression is required to determine sensory axon trajectory, branching, and target choice. *J. Neurosci.* **22,** 832–841.

McConnell, S. K., Ghosh, A., and Shatz, C. J. (1989). Subplate neurons pioneer the first axon pathway from the cerebral cortex. *Science* **245,** 978–982.

Miech, C., Pauer, H. U., He, X., and Schwarz, T. L. (2008). Presynaptic local signaling by a canonical wingless pathway regulates development of the *Drosophila* neuromuscular junction. *J. Neurosci.* **28,** 10875–10884.

Miller, D. M., Shen, M. M., Shamu, C. E., Burglin, T. R., Ruvkun, G., Dubois, M. L., Ghee, M., and Wilson, L. (1992). *C. elegans* unc-4 gene encodes a homeodomain protein that determines the pattern of synaptic input to specific motor neurons. *Nature* **355,** 841–845.

Mitchell, K. J., Doyle, J. L., Serafini, T., Kennedy, T. E., Tessier-Lavigne, M., Goodman, C. S., and Dickson, B. J. (1996). Genetic analysis of Netrin genes in *Drosophila*: Netrins guide CNS commissural axons and peripheral motor axons. *Neuron* **17,** 203–215.

Mori, I., and Ohshima, Y. (1995). Neural regulation of thermotaxis in *Caenorhabditis elegans*. *Nature* **376,** 344–348.

Murai, K. K., and Pasquale, E. B. (2004). Eph receptors, ephrins, and synaptic function. *Neuroscientist* **10,** 304–314.

O'Leary, D. D., Chou, S. J., and Sahara, S. (2007). Area patterning of the mammalian cortex. *Neuron* **56,** 252–269.

Peteya, D. J. (1973). A light and electron microscope study of the nervous system of *Ceriantheopsis americanus* (Cnidaria, Ceriantharia). *Z. Zellforsch Mikrosk Anat.* **141,** 301–317.

Plachez, C., and Richards, L. J. (2005). Mechanisms of axon guidance in the developing nervous system. *Curr. Top. Dev. Biol.* **69,** 267–346.

Polleux, F., Ince-Dunn, G., and Ghosh, A. (2007). Transcriptional regulation of vertebrate axon guidance and synapse formation. *Nat. Rev. Neurosci.* **8,** 331–340.

Poon, V. Y., Klassen, M. P., and Shen, K. (2008). UNC-6/netrin and its receptor UNC-5 locally exclude presynaptic components from dendrites. *Nature* **455,** 669–673.

Rand, J. B., and Nonet, M. L. (1997). *C. elegans* II. *In* "Synaptic Transmission" (D. L. Riddle, T. Blumenthal, B. J. Meyer, and J. R. Priess, Eds.), vol. II, pp. 611–643. Cold Spring Harbor Laboratory Press, Cold Spring Harbor.

Sabo, S. L., Gomes, R. A., and McAllister, A. K. (2006). Formation of presynaptic terminals at predefined sites along axons. *J. Neurosci.* **26,** 10813–10825.

Sakarya, O., Armstrong, K. A., Adamska, M., Adamski, M., Wang, I. F., Tidor, B., Degnan, B. M., Oakley, T. H., and Kosik, K. S. (2007). A post-synaptic scaffold at the origin of the animal kingdom. *PLoS ONE* **2,** e506.

Salie, R., Niederkofler, V., and Arber, S. (2005). Patterning molecules; multitasking in the nervous system. *Neuron* **45,** 189–192.

Salinas, P. C. (2005). Signaling at the vertebrate synapse: New roles for embryonic morphogens? *J. Neurobiol.* **64,** 435–445.

Salinas, P. C., and Zou, Y. (2008). Wnt signaling in neural circuit assembly. *Annu. Rev. Neurosci.* **31,** 339–358.

Scheiffele, P. (2003). Cell-cell signaling during synapse formation in the CNS. *Annu. Rev. Neurosci.* **26,** 485–508.

Shen, K., and Bargmann, C. I. (2003). The immunoglobulin superfamily protein SYG-1 determines the location of specific synapses in *C. elegans*. *Cell* **112,** 619–630.

Shen, K., Fetter, R. D., and Bargmann, C. I. (2004). Synaptic specificity is generated by the synaptic guidepost protein SYG-2 and its receptor, SYG-1. *Cell* **116,** 869–881.

Shishido, E., Takeichi, M., and Nose, A. (1998). Drosophila synapse formation: Regulation by transmembrane protein with Leu-rich repeats, CAPRICIOUS. *Science* **280,** 2118–2121.

Simon, D. J., Madison, J. M., Conery, A. L., Thompson-Peer, K. L., Soskis, M., Ruvkun, G. B., Kaplan, J. M., and Kim, J. K. (2008). The microRNA miR-1 regulates a MEF-2-dependent retrograde signal at neuromuscular junctions. *Cell* **133,** 903–915.

Sudhof, T. C. (2004). The synaptic vesicle cycle. *Annu. Rev. Neurosci.* **27,** 509–547.

Tessier-Lavigne, M., and Goodman, C. S. (1996). The molecular biology of axon guidance. *Science* **274,** 1123–1133.

Vaughn, J. E. (1989). Fine structure of synaptogenesis in the vertebrate central nervous system. *Synapse* **3,** 255–285.

Von Stetina, S. E., Fox, R. M., Watkins, K. L., Starich, T. A., Shaw, J. E., and Miller, D. M. 3rd. (2007). UNC-4 represses CEH-12/HB9 to specify synaptic inputs to VA motor neurons in *C. elegans*. *Genes Dev.* **21** 332–346.

Von Stetina, S. E., Treinin, M., and Miller, D. M. 3rd. (2006). The motor circuit. *Int. Rev. Neurobiol.* **69** 125–167.

Waites, C. L., Craig, A. M., and Garner, C. C. (2005). Mechanisms of vertebrate synaptogenesis. *Annu. Rev. Neurosci.* **28,** 251–274.

Westfall, I. A. (1996). Ultrastructure of synapses in the first-evolved nervous systems. *J. Neurocytol.* **25,** 735–746.

White, J. G., Southgate, E., and Thomson, J. N. (1992). Mutations in the *Caenorhabditis elegans* unc-4 gene alter the synaptic input to ventral cord motor neurons. *Nature* **355,** 838–841.

White, J. G., Southgate, E., Thomson, J. N., and Brenner, S. (1986). The structure of the nervous system of the nematode *Caenorhabditis elegans*. *Philosophical Transactions of the Royal Society of London* **314,** 1–340.

Yamagata, M., Sanes, J. R., and Weiner, J. A. (2003). Synaptic adhesion molecules. *Curr. Opin. Cell. Biol.* **15,** 621–632.

Yamagata, M., Weiner, J. A., and Sanes, J. R. (2002). Sidekicks: Synaptic adhesion molecules that promote lamina-specific connectivity in the retina. *Cell* **110,** 649–660.

CHAPTER THREE

The Developmental Integration of Cortical Interneurons into a Functional Network

Renata Batista-Brito*,†,‡ *and* Gord Fishell*,†

Contents

* Smilow Neuroscience Program, New York University Medical Center, New York, USA
† Department of Cell Biology, New York University Medical Center, New York, USA
‡ Gulbenkian PhD Programme in Biomedicine, Gulbenkian Science Institute, Oeiras, Portugal

Current Topics in Developmental Biology, Volume 87
ISSN 0070-2153, DOI: 10.1016/S0070-2153(09)01203-4

Abstract

The central goal of this manuscript is to survey our present knowledge of how cortical interneuron subtypes are generated. To achieve this, we will first define what is meant by subtype diversity. To this end, we begin by considering the mature properties that differentiate between the different populations of cortical interneurons. This requires us to address the difficulties involved in determining which characteristics allow particular interneurons to be assigned to distinct subclasses. Having grappled with this thorny issue, we will then proceed to review the progressive events in development involved in the generation of interneuron diversity. Starting with their origin and specification within the subpallium, we will follow them up through the first postnatal weeks during their integration into a functional network. Finally, we will conclude by calling the readers attention to the devastating consequences that result from developmental failures in the formation of inhibitory circuits within the cortex.

1. General Introduction

Alber Szent-Gyogyi once said: "If structure does not tell us anything about function, it only means that we have not looked at it correctly" (Buzsaki, *Rhythms of the Brain*, 2006). A century ago, our understanding of brain structure was limited to descriptions of its gross anatomy. Cajal's recognition that the neuron was the fundamental building block of the brain provided the first meaningful step in the exploration of how brain structure at a cellular level relates to its ability to manifest complex behaviors. Since then, great strides have been made in unraveling how the topographic projections of excitatory pyramidal neurons and associated inhibitory interneurons contribute to the creation of sensory representations of the physical world. However, less progress has been made in understanding the learning and memory consolidation within the neocortex. Cortical interneurons, with their enormous diversity of subtypes, and their ability to provide feedforward and feedback information are ideally suited for this role. Each interneuron subclass likely possesses both specific connectivity and an input/output function that define their role in the cortex. Here we explore the logic by which genetic and environmental events contribute to the generation of cortical interneuron subtypes and the events leading to their subsequent integration into cortical networks.

2. The Classification of Cortical Interneuron Subtypes

Recently, a consortium of scientists specializing in development, anatomy and/or physiology convened at the birthplace of Ramon y Cajal for 3 days to discuss interneuron classification (PING, 2008). Adding to the historical spirit of this venture, the meeting was held in the fifteenth-century chapel where in his youth Cajal was no doubt compelled to languish. This committee vetted itself with the task of creating a unifying nomenclature for GABAergic interneurons in the cerebral cortex, which they ultimately designated as the Petilla terminology (Petilla Interneuron Nomenclature Group *et al.*, 2008). Following the recommendations of this committee, we will review the different characteristics attributed to distinct subtypes of interneurons in the mature cortex. This includes consideration of the characteristic morphologies, molecular markers, and physiological properties that characterize distinct mature cortical interneuron populations.

2.1. What is an interneuron? How many subtypes? Basic properties of mature interneurons

GABAergic cortical interneuron diversity is achieved by the acquisition of different properties such as morphology, molecular, targeting, synaptic transmission and physiological character. On the basis of these traits, cortical interneurons have been subdivided into different subclasses. Ideally, all the criteria would be taken into account in determining the specific designation of particular interneurons; however, the difficulty of performing parametric analysis makes this in most cases impractical. Moreover, as scientists often disagree on the appropriate weighting of specific characteristics in classifying these cells, an interneuron's subtype is all too often left in the eye of the beholder.

Interneurons were first identified and classified based on their morphology. In 1899, Cajal described them as "short axon cells," whose axons project locally (however, if stretched out these axons are anything but short). Axon morphology has been historically considered to be a critical characteristic for determining interneuron subtype (with the position of the initial segment, arbor trajectory, branch metrics, and synaptic subcellular selectivity typically being used as classifiers). Emphasis on this feature seems sensible, as the extent of an axon's ramifications will to a great extent determine its output domain. A horizontal or vertical arbor will establish if the output will influence individual or multiple cortical columns or layers. Dendrites (arborization polarity and branch metrics) and the morphology (shape and size) of the cell soma are also used for classification (Petilla Interneuron Nomenclature Group *et al.*, 2008). Another important criteria to consider is the precise cell types a particular interneuron targets, such as pyramidal cells, other interneurons, glial cells, or cells of the vascular system, as well as the subcellular compartment (somata, dendrites, axon, etc.) that they selectively target (Petilla Interneuron Nomenclature Group *et al.*, 2008).

In addition, molecular markers that account for many of the neuron's emergent properties, such as neuropeptides, Ca^{2+} proteins, ionic channels, receptors, and transporters have been typically used for interneuron classification, (Baraban and Tallent, 2004; Goldberg *et al.*, 2005, 2008; Lau *et al.*, 2000; Monyer and Markram, 2004; Rudy and McBain, 2001). By combining gene expression techniques with intracellular recording, dye filling, and morphological analysis, it has been shown that some molecular markers such as parvalbumin (PV), calretinin (CR), Kv3.1, vasoactive intestinal peptide (VIP), somatostatin (SST), cholecystokinin (CCK), and neuronal nitric oxide synthase (nNOS) are good indicators of subtype identity, while others, such as calbindin (CB), neuropeptide Y (NPY), and Kv3.2, are more difficult to use due to their expression by a variety of cell types (Cauli *et al.*, 1997; Chow *et al.*, 1999; DeFelipe, 1993; Gerashchenko *et al.*, 2008; Gonchar and Burkhalter, 1997; Gupta *et al.*, 2000; Kawaguchi and Kubota, 1996; Kubota and Kawaguchi, 1994, 1997; Monyer and Markram, 2004). Comprehensive subtype analysis has become possible with the advent of microarray technology. Genome-wide profiling of small groups of interneurons (Sugino *et al.*, 2006) or even single cells (Kamme *et al.*, 2003) is becoming increasingly common.

A further criterion widely utilized for interneuron classification is their intrinsic firing properties, which are indicative of the activity of particular neurons and their presumed role within the cortical circuitry. Different classes of interneurons have distinct firing and subthreshold properties. For example, in response to a long depolarizing current pulse, interneurons show a variety of distinct firing characteristics, including bursting, stuttering, fast spiking, irregular spiking, and accomodation (Petilla Interneuron Nomenclature Group *et al.*, 2008). Some caution, however, must be applied to the interpretation of these depolarization patterns, as they may not occur under physiological conditions. It is difficult to gauge the significance of a neuron stuttering when it normally only speaks in short sentences.

Interneurons are also distinguishable at the synaptic level, by their inputs, the target specificity of their output and postsynaptic responses (Somogyi and Klausberger, 2005; Somogyi *et al.*, 1998). The response of interneurons to excitatory or inhibitory input can vary depending on the receptor subtype mediating these responses. For example, axoaxonic interneurons (or chandelier cells) despite being inhibitory produce a postsynaptic GABAergic depolarization of the initial segment of pyramidal cells. Such postsynaptic selectivity was recently shown to have dramatic functional consequences at the microcircuit level in humans (Molnar *et al.*, 2008). Moreover, a neuron's prior history of stimulation can affect how it responds to later synaptic events.

As a consequence, repeated firing can result in facilitation or depression (Gupta *et al.*, 2000). The intrinsic and synaptic properties are both likely a product of the molecular composition of the cell such as channels and other signaling molecules, as well as the ramification of their processes. Therefore it is not surprising that in many cases there is a correlation between the intrinsic electrophysiology and synaptic responses of an interneuron, and their molecular markers and morphology. However, this correlation is not absolute, and one cannot *a priori* predict the functional characteristics of an interneuron solely on the basis of their molecular signature. Thus, to both classify and assess the function of a certain interneuron subtype, it is essential to take a multifaceted approach, including as many of the features referred above (and summarized in Table 3.1) as possible.

The so-called fast spiking, basket cells provide a good example of this sort of taxonomy in practice. These cells are primarily PV expressing and are classified by their innervation at the soma and proximal dendrite of target neurons (both pyramidal and GABAergic) (Freund, 2003). Among their archtypical characteristics are their fast kinetics, including their intrinsic firing (high-frequency repetitive firing, brief single spikes, a fast membrane time constant) and their synaptic outputs and inputs (Jonas *et al.*, 2004). These to a large degree are accounted for by their expression of specific channels, including voltage-gated potassium (K^+) channels (namely Kv3.1), sodium (Na^+) channels with rapid activation kinetics, and AMPA receptors with fast gating kinetics (Jonas *et al.*, 2004; Rudy and McBain, 2001). They innervate postsynaptic targets close to the site of action potential initiation, the axon initial segment. As a result they have been shown to provide strong inhibitory control over postsynaptic cell firing and are the main class of cells to provide feedforward inhibition and hence precise control of neuronal output (Pouille and Scanziani, 2001). Moreover, as a given PV basket cell contacts a vast number postsynaptic cells, this interneuron subtype has the capacity to synchronize the firing of large groups of neurons (Cobb *et al.*, 1995; Pouille and Scanziani, 2001). Although we have chosen the FS basket cells to illustrate the mature properties of a particular interneuron subtype equally intriguing characteristics have been ascribed to many of the other major classes of cortical interneurons. Recent literature provides detailed discussions of some of the more prominent subclasses, including VIP-expressing multipolar, CR-expressing bipolar, or Sst-expressing Martinotti interneuron subclasses (Ferezou *et al.*, 2002; Ma *et al.*, 2006; Markram *et al.*, 2004; Porter *et al.*, 1998).

Despite the seductive wealth of data describing subtype complexity, in the final analysis the plaintive warning given by Dr Somogyi at the beginning of the Petilla meeting should be heeded:

> The shapes, markers and firing patterns of interneurons exist as separate entities only in the collective imagination of investigators. In the end there is only the contextual input-output function of interneurons, as they

Table 3.1 Summary of morphological, molecular, and physiological features (from the Petilla meeting)

Morphological features
Axon: initial segment, arbor trajectory, terminal shape, branch metrics, boutons, synaptic targets
Dendrite: arborization polarity, branch metrics, fine structure, postsynaptic element
Soma: shape, size, orientation
Connections: chemical and electrical, source, location and distribution
Molecular features
Transcription factors
Neurotransmitters or their synthesizing enzymes
Neuropeptides
Calcium-binding proteins
Receptors: ionotropic, metabotropic
Structural proteins
Cell-surface markers
Ion channels
Connexions
Transporters: plasma membrane, vesicular
Physiological features
Passive or subthreshold parameters: resting membrane potential, membrane time constants, input resistance, oscillation and resonance, rheobase and chronaxie, rectification
Action potential (AP) measurements: amplitude, threshold, half-width, afterhyperpolarization, afterdepolarization, changes in AP waveform during train
Dendritic backpropagation
Depolarizing plateaus
Firing pattern: oscillatory and resonant behavior, onset response to depolarizing step, steady-state response to depolarizing step
Response to hyperpolarizing step: rectification, rebound
Spiking recorded extracellularly: phase relationship to oscillations, functional response specificity, crosscorrelation and other dynamics
Postsynaptic responses: spontaneous and evoked, ratio of receptor subtypes, spatial and temporal summation, short- and long-term plasticity, gap junctions

perform operations via electrical and chemical signals. Thus, their reality is only tangible to the extent that they provide effective neural modulation to a functioning network, as it transits through different brain states. (P. Somogyi, personal communication).

3. Development of Cortical Interneurons

Inhibitory (GABAergic) interneurons, although far less numerous than the excitatory pyramidal neurons, are comprised by at least 15 subtypes, (Markram *et al.*, 2004; Petilla Interneuron Nomenclature Group *et al.*, 2008). This large variety in inhibition is thought to bestow the brain with the profound nonlinear functionality required for higher cortical processes. While to date no systematic studies have been done across various species, we predict that such a study would reveal that the diversity of interneurons compared to projection neurons has increased disproportionately during speciation. Furthermore, as we believe that much of this diversity is determined during development, examining the developmental origins of cortical interneurons should ultimately prove to be highly informative. Here, we survey the current literature and provide our present view of how genetic and environmental factors result in the generation of different cortical interneuron subtypes, as well as their subsequent integration into cortical circuitry.

3.1. The telencephalon is an alar plate derivative

Comparative fate-mapping studies in a variety of organisms indicate that the telencephalon is derived from the anterior lateral (i.e., alar) neural plate, and subdivided into specific regional territories defined by morphogenetic parameters during development (Fishell, 1997; Rubenstein *et al.*, 1998; Shimamura *et al.*, 1995, 1997). Pioneering efforts to understand the compartmentalization of the telencephalon have lead to the proposal of the prosomeric model (Pombal and Puelles, 1999; Puelles, 2001; Puelles and Rubenstein, 2003; Rubenstein *et al.*, 1994). This model posits that early lineage restrictions result in the establishment of transient prosomeres (defined as segment-like divisions) within the forebrain (Rubenstein *et al.*, 1994). The existence and exact location of the prosomeric boundaries are still a matter of debate; however, some of the divisions of the telencephalon are today universally accepted. The most prominent subdivision separates the dorsal pallium and ventral subpallium. Rather than being a result of lineage restriction, early dorsoventral patterning events appear to be mediated by extrinsic cues (for reviews, see Hérbert and Fishell, 2008; Rallu *et al.*, 2002a; Sur and Rubenstein, 2005; Wilson and Houart, 2004; Wilson and Rubenstein, 2000).

Recent studies suggest that BMPs and Wnts bestow the pallium with dorsal character (Aboitiz and Montiel, 2007; O'Leary and Sahara, 2008; O'Leary *et al.*, 2007). Glutamatergic cortical cells (or principal cells) originate within the ventricular and subventricular region of the pallium and migrate radially to their final destination (Kriegstein and Noctor, 2004; Marin and Rubenstein, 2003; Mission *et al.*, 1991; Parnavelas, 2000; Rakic, 1990). By contrast, the ventral telencephalon (subpallium) responds to Shh and FGF8 by producing different types of GABAergic neurons, such as cortical interneurons and striatal medium spiny neurons. Postmitotic interneurons of all mammals studied to date (and probably many vertebrates) undergo long-range tangential migration to reach the cortex (Corbin *et al.*, 2001; Kriegstein and Noctor, 2004; Marin and Rubenstein, 2001, 2003). Surprisingly, a possible exception to this occurs in humans, where it has been reported that interneurons undergo at least their final mitotic events in the subventricular zone (SVZ) of the pallium (Letinic *et al.*, 2002).

3.2. The embryonic subpallium: From embryology to developmental biology

The ganglionic eminences were first described as bumps of proliferating cells within the ventral telencephalon. Later, it was discovered that these regions could be subdivided based on their expression of specific genes. Loss of function analysis of these genes, as well as fate-mapping efforts showed that these regions produce different types of neurons. The next section follows along historical lines to consider (1) the embryology, (2) molecular specification, and (3) *in vivo* fate mapping of the subpallium.

3.3. The ganglionic eminences were first defined by their morphology

Landmark studies by Smart and Sturrock (Smart, 1976; Sturrock and Smart, 1980) subdivided the subpallium into discrete territories based on anatomical features within the neuroepithelium. They described the murine lateral (LGE) and medial (MGE) ganglionic eminences by analogy to what was first described in human embryos (Smart, 1976; Sturrock and Smart, 1980). From the beginning, it was appreciated that these ganglionic eminences arose in a discrete temporal order with the "earlier-generated ventral eminence" (the MGE) appearing at the telencephalic–diencephalic junction, followed by a "later appearing dorsal eminence," named the lateral ganglionic eminence (LGE) (Smart, 1976). The caudal ganglionic eminence (CGE) was proposed much later as a discrete entity being first defined by Anderson *et al.* (2001) as the eminence that is posterior to the fusion of the MGE and LGE. However, while the MGE and LGE are clearly separated by a sulcus, a physical barrier does not exist in the case of the CGE (see figure 3.1), and therefore its identity based on embryology is controversial, a matter that we will address later. Another progenitor region within the subpallium that has received little attention to date is the septal eminence. Like the CGE, this structure arises relatively late in development. While the septal eminence

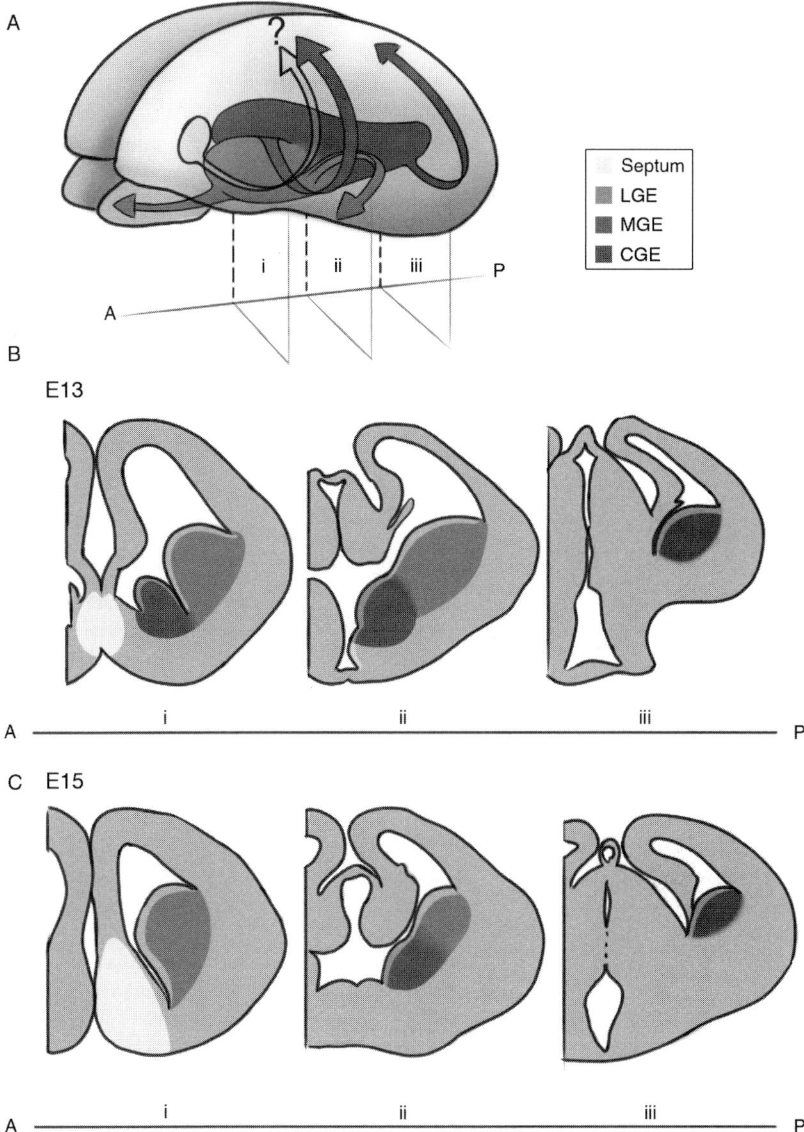

Figure 3.1 Schematic representation of the developing murine brain. (A) Three-dimensional view of a developing brain indicating the ventral anatomical regions that give rise to interneurons and their correspondent migratory pathways. The LGE (green) generates interneurons that migrate to the olfactory bulb and striatum. Both the MGE (red) and CGE (blue) produce cortical interneurons. Whether the septum produces cortical interneurons (yellow) is still a matter of debate. (B–C) Coronal views of the brain in (A) at three locations (i–iii) along the anterior (A)–posterior (P) axis at the embryonic ages E13 (B) and E15 (C). (See Color Insert.)

has been assumed to give rise to the septal nucleus, this structure has yet to be fate mapped. Although studies are too preliminary for a definitive answer, hints exist that this eminence may also be a source for tangentially migrating cell populations, which possibly include a subset of cortical interneurons (Taglialatela *et al.*, 2004; Xu *et al.*, 2008). In addition, less well morphologically defined regions such as the anterior entopeduncular region (AEP) and the preoptic area (POa) may also give rise to unique populations, most likely within the amygdala complex (Bulfone *et al.*, 1993; Cobos *et al.*, 2001; Flames *et al.*, 2007; Puelles *et al.*, 2000) (Fig. 3.1).

3.4. The ganglionic eminences as defined by molecules

The subpallium undergoes gradual morphological changes during development. This complicates the identification of the eminences solely on the basis of anatomical features. To overcome this limitation there has been a major effort, spearheaded by the Rubenstein laboratory, to discover the molecular markers that correspond with these anatomical differences. In this section, we will present what is currently known about the molecular and genetic mechanisms that generate cell type diversity within the ventral eminences, and discuss insights regarding how specific transcription factors might regulate cell fate decisions.

Many genes are broadly expressed within the three eminences and are known to be essential for the generation of GABAergic cells, as well as more specifically for cortical interneuron development. In particular, members of the *Dlx* gene family, which are expressed throughout the subpallial SVZ, have been shown to be critical for interneuron specification (Anderson *et al.*, 1997b; Petryniak *et al.*, 2007; Pleasure *et al.*, 2000). Mice containing compound *Dlx1/Dlx2* mutations die at birth and have a severe reduction in the tangential migration of interneurons from the ventral eminences to the neocortex, resulting in a massive loss of neocortical GABAergic cells at birth (Anderson *et al.*, 1997a). Furthermore, *Dlx1/Dlx2* nulls have abnormal striatal differentiation (Anderson *et al.*, 1997b); and virtually no olfactory bulb interneurons (Bulfone *et al.*, 1998). Similarly, null mutations in *Mash1*, a proneural gene expressed throughout the subpallial SVZ (Porteus *et al.*, 1994) display a marked loss of GABAergic cortical and olfactory bulb interneurons (Casarosa *et al.*, 1999).

While *Dlx* family and *Mash1* genes are broadly expressed within the subpallium, other genes show more restricted expression patterns. Loss of function of such these genes revealed that ganglionic subdivisions generate different subtypes of interneurons. It seems likely that with the discovery of novel gene expression patterns, finer graded subdivisions within the eminences will be established.

3.4.1. Molecular development of the LGE

Gene expression analyses coupled with loss of function studies have shown that the LGE, which broadly expresses *Pax6* and *Gsh2*, can be subdivided into a dorsal and ventral domain (Flames *et al.*, 2007; Yun *et al.*, 2001). The dorsal domain of the LGE (dLGE) anatomically separates the pallium and the subpallium. This region expresses genes characteristic of the pallium such as *Pax6*, *Ngn2*, and *Dbx1* (Flames *et al.*, 2007; Puelles *et al.*, 2000; Yun *et al.*, 2001), and it gives rise to the postnatal SVZ, as well as to olfactory bulb interneurons (Stenman *et al.*, 2003a). The ventral LGE (vLGE) expresses *Gsh2*, and low levels of *Pax6*, and primarily gives rise to the striatum (Flames *et al.*, 2007; Puelles *et al.*, 2000; Stenman *et al.*, 2003a; Waclaw *et al.*, 2006). In fact, most striatal projection neurons derive from the LGE (Anderson *et al.*, 1997b; Deacon *et al.*, 1994; Olsson *et al.*, 1995, 1998; Stenman *et al.*, 2003a; Wichterle *et al.*, 2001).

Gsh genes are required for LGE development. In *Gsh1/2* mutants, the LGE becomes dorsalized, and shows reduced expression of *Mash1* and *Dlx* genes. By contrast, the MGE (which expresses both of these genes, albeit in the case of *Gsh2*, somewhat later) does not appear to be affected (Corbin *et al.*, 2000; Toresson and Campbell, 2001; Toresson *et al.*, 2000; Yun *et al.*, 2001, 2003). During development, *Er81* is restricted to the dLGE (also note that a more ventral region of *Er81* expression is observed within the MGE) (Stenman *et al.*, 2003a). Suggesting their genetic hierarchy, in *Gsh1/2* mutants the expression of *Er81* in the dLGE is lost. Similarly, *Sp8* is expressed within the dLGE, in the postnatal SVZ, the rostral migratory stream (RMS), and eventually in the calretinin (CR) olfactory bulb interneurons arising from this progenitor zone. Conditional removal of this gene within the *Dlx5/6* domain results in a decrease of the CR olfactory bulb interneuron subtype (Waclaw *et al.*, 2006). Taken together, the LGE can be subdivided into a ventral part that produces projection neurons and a dorsal portion that generates interneurons. Notably, although the dLGE appears to be the primary source of interneurons for the olfactory bulb, recent analysis has indicated subsets of this population are generated within both the pallium and the septum (Merkle *et al.*, 2007) (Fig. 3.2).

3.4.2. Molecular development of the MGE

The majority of the MGE expresses *Nkx2.1* (Sussel *et al.*, 1999). By contrast, the dorsal MGE expresses *Nkx6.2* and *Gli1* and is partially *Nkx2.1* negative (Fogarty *et al.*, 2007; Rallu *et al.*, 2002b; Wonders *et al.*, 2008). *Nkx2.1* null mice have an atrophied MGE and show a drastic reduction of cortical interneurons at birth (Sussel *et al.*, 1999). *Nkx2.1* is required for the correct specification of MGE-derived interneuron subtypes and for the establishment and maintenance of MGE identity (Butt *et al.*, 2008). Conditional removal of *Nkx2.1* leads to an interneuron subtype fate

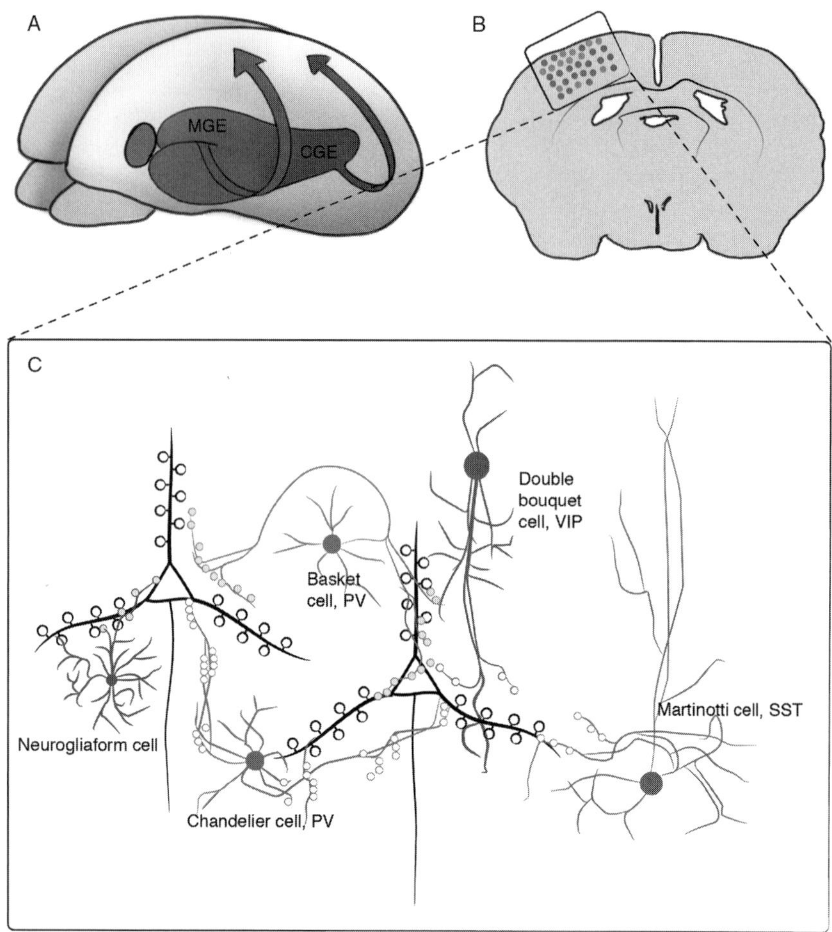

Figure 3.2 Differential origin of cortical interneuron subtypes. (A) Three-dimensional view of an embryonic murine brain highlighting the two ventral regions that produce cortical interneurons, the MGE (red) and the CGE (blue). (B) Coronal view of an adult brain, illustrating the proportion and relative distribution of cortical interneurons derived from the MGE (red dots) and the CGE (blue dots). (C) Detailed schematic view of the boxed region in (B) illustrating the main interneuron subtypes derived from the MGE (red cells) and CGE (blue cells), and how they characteristically interact with the pyramidal cells (black cells). This figure is adapted from Kawaguchi and Kubota (2002). (See Color Insert.)

switch in a time-dependent manner. Similar to what was observed in *Nkx2.1* nulls (Sussel *et al.*, 1999), early conditional removal (E10.5) of *Nkx2.1* leads to a decrease of cortical interneurons and a corresponding increase in striatal spiny neurons (Butt *et al.*, 2008). It also results in the remaining cortical interneurons acquiring characteristics normally found in the CGE population. Later removal of *Nkx2.1* at E12.5 resulted in a decrease of MGE-derived PV/SST cortical interneuron subtypes, and an increase in the normally CGE-derived CR/VIP population.

It has recently been shown that *Nkx2.1* is upstream of the LIM-homeobox transcription factor *Lhx6* and specifies interneuron fate by directly activating this gene (Du *et al.*, 2008). MGE-derived progenitors start to express *Lhx6* as soon as they leave the ventricular zone (Grigoriou *et al.*, 1998). *Lhx6* expression persists through adulthood in most parvalbumin (PV)- and somatostatin (SST)-expressing cortical interneurons (Cobos *et al.*, 2005; Du *et al.*, 2008; Fogarty *et al.*, 2007; Gong *et al.*, 2003; Lavdas *et al.*, 1999; Liodis *et al.*, 2007). *Lhx6* loss of function analysis has shown that this gene is required for the normal specification and migration of MGE-derived GABAergic cells, and null animals exhibit a loss of PV and SST interneurons in the neocortex and hippocampus (Liodis *et al.*, 2007). Hence, the MGE appears to be the sole source of three of the most prominent interneuron subtypes the basket, chandelier, and Martinotti cells.

3.4.3. Molecular development of the CGE is a black box

In addition to the MGE, the CGE is the other main source of cortical interneurons. Although our initial analysis at E13.5 suggested that the CGE accounted for 15% the total cortical interneuron population (Nery *et al.*, 2002), recent results from our group suggests that its contribution may actually be far larger, and contribute as much as 40% of all cortical interneurons (G. Miyoshi *et al.*, unpublished results). To date, no CGE specific transcription factors have been identified. However, it has been recently shown that CGE-derived interneurons specifically express the serotonin receptor *5HT3a*, while *Nkx6.2* and *CoupTF1/2* are widely but not selectively expressed within the CGE (Sousa *et al.*, 2009). The similarity of genes expressed in the CGE and dMGE (namely *Nkx6.2* and *CoupTF1/2*) raises the possibility that they might be considered a common structure. Accordingly, in our *Nkx2.1* loss of function analysis, we observed the expansion of both CGE and dMGE gene expression (Sousa *et al.*, 2009). Alternatively, emerging fate-mapping studies suggest that the dMGE appears to have a hybrid identity producing both interneuron subtypes characteristic of the MGE and CGE (Fogarty *et al.*, 2007). In summary, while the CGE is a distinct anatomical progenitor region that originates cortical interneurons, no specific molecular identity has been attributed to this structure, making its distinction a difficult task.

3.4.4. The undiscovered country

In addition, as noted above it is possible that the septum, AEP and POa may also contribute to the production of cortical interneurons (Bulfone *et al.*, 1993; Cobos *et al.*, 2001; Flames *et al.*, 2007; Puelles *et al.*, 2000). These structures express genes characteristic of the ganglionic eminences, such as

Nkx2.1, *Dlx1/2*, and *Vax1* (Puelles *et al.*, 2000; Taglialatela *et al.*, 2004). *Vax1* null mice have a reduced MGE and virtually no septum, and show a loss of about 40% of cortical interneurons at birth (Taglialatela *et al.*, 2004). Selective fate mapping of the septum will be necessary to definitively address this issue. It seems likely that further analysis of these structures will produce a few more surprises, such as the recent realization that the AEP generates interneurons destined to specific amygdala nuclei (Hirata *et al.*, 2009).

3.5. Fate mapping of the ganglionic eminences

Gene expression and loss of function analyses have given us strong indications about the differential production of cortical interneuron subtypes within specific progenitor domains. However, the most direct way of addressing this issue is by selective fate mapping of these embryonic regions. In this section, we will summarize the results of different fate-mapping approaches that have been undertaken over the last decade. As will be evident, these findings both confirm and extend the fate conclusions based on expression and loss of function analyses.

Transplantation of labeled progenitors into the different eminences has shown that the MGE and LGE produce both GABAergic projection neurons (such as the spiny striatal neurons) and interneurons (cortical interneurons from the MGE and olfactory bulb interneurons from the LGE) (Nery *et al.*, 2002; Wichterle *et al.*, 2001; Xu *et al.*, 2004); while the CGE produces only interneurons (Nery *et al.*, 2002). Multiple studies have demonstrated that interneurons born in the MGE and CGE migrate tangentially to enter the cortex, while interneurons born in the LGE migrate via the RMS to reach the olfactory bulb (Batista-Brito *et al.*, 2008a; Corbin *et al.*, 2001; Marin and Rubenstein, 2003). Based on tracing experiments and the migration patterns of cells from the eminences into the cortex, it was initially thought that the LGE was a primary source of cortical interneurons (de Carlos *et al.*, 1996; Tamamaki *et al.*, 1997). While the possibility that the LGE produces a small subpopulation of cortical interneurons has not been entirely ruled out, this conclusion can be largely explained by the fact that MGE- and CGE-derived interneurons migrate through the LGE while en route to the cortex (Anderson *et al.*, 2001; Wichterle *et al.*, 1999, 2001).

Fate mapping of the MGE has shown that a number of distinct interneuronal populations are generated in this region and subsequently migrate tangentially to populate distant telencephalic regions, such as the striatum, cortex, and hippocampus (Anderson *et al.*, 2001; Lavdas *et al.*, 1999; Wichterle *et al.*, 1999, 2001). Transplants of labeled MGE progenitors *in vivo* (Butt *et al.*, 2005; Wichterle *et al.*, 2001) and *in vitro* experiments (Xu *et al.*, 2004) have confirmed that this region produces PV- and SST-expressing interneuron subtypes that account for the majority of interneurons in the cortex (see figure 3.2). In terms of their intrinsic electrophysiological properties, the most abundant interneuron subtype produced in the MGE is the fast-spiking cell (FS), while the second largest population is the burst-spiking Martinotti interneurons (BSNP) (Butt *et al.*, 2005). In addition, regular-spiking nonpyramidal (RSNP although this particular nomenclature is likely to be soon supplanted by more refined physiological criteria, Miyoshi *et al.* unpublished) that are largely NPY-positive interneurons are also derived from this region, but some arise from the CGE as well.

Recently, Cre/loxP-mediated genetic fate-mapping experiments have confirmed and extended previous transplantation results by providing selective labeling of MGE and CGE lineages. By using different Cre drivers of genes expressed in the MGE such as *Nkx2.1*, *Nkx6.2*, and *Lhx6* (Fogarty *et al.*, 2007; Miyoshi *et al.*, 2007; Xu *et al.*, 2008) it has been possible to address some of the regional aspects of MGE-derived interneuron diversity. In concordance with the transplantation studies from the MGE, genetic fate mapping of this structure by using the allele Nkx2.1^{Cre} shows that this region produces primarily PV, SST, and CB, and to a lesser extent NPY and CR neurons (Fogarty *et al.*, 2007; Xu *et al.*, 2008). Genetic fate mapping of Nkx6.2 progenitors (expressed in the sulcus) using an Nkx6.2^{Cre} allele shows that this region, similar to the *Nkx2.1* domain, produces CB, SST, PV, NPY, and CR neurons (Fogarty *et al.*, 2007; Stenman *et al.*, 2003b). However, fate mapping of the Nkx6.2 region yields a proportionally greater contribution to the NPY and CR subtypes than the *Nkx2.1* domain fate map, suggesting that the latter two subtypes preferentially originate in the most dorsal part of the MGE (Fogarty *et al.*, 2007; Xu *et al.*, 2008). Fate mapping of Lhx6-expressing cells reveals a total overlap of the Lhx6 lineage with the cortical PV, SST, and CB interneurons (Fogarty *et al.*, 2007; Xu *et al.*, 2008), and the contribution of Lhx6 for NPY and CR subtypes is similar to that observed for Nkx6.2 (Fogarty *et al.*, 2007; Xu *et al.*, 2008).

Transplantation experiments have found that the E13.5 CGE contributes interneurons to different regions of both the dorsal and ventral telencephalon, such as the cortex, specific regions of the limbic system (amygdala, hippocampus, and NAc) and the striatum (Nery *et al.*, 2002). However, in contrast to the MGE and LGE, the CGE transplants produced few GABAergic projection neurons (Nery *et al.*, 2002). This is likely due to the fact that this eminence only develops relatively late in embryogenesis, when the switch from the production of ventral GABAergic projection neurons to cortical interneurons has already occurred. Homotopic transplants of E13.5 and E15.5 CGE progenitors have shown that the majority of the cortical cells from this region are either CR-, NPY-, or VIP-expressing interneurons (see figure 3.2). These cells have mainly bipolar or multipolar morphologies, and are RSNPs and NR-RSNPs in terms of their electrophysiological profile (Butt *et al.*, 2005). Due to the fact that we still do not know of any gene that specifically labels the CGE, genetic fate mapping of this eminence has been hampered. Recently, the fortuitous expression of a

BAC Mash1^{CreER} transgenic line selectively in the CGE and not MGE, has allowed for a detailed genetic fate-mapping analysis of this region. This study in progress has confirmed the results from previous CGE transplant experiments, and has expanded upon them by indicating that the contribution of CGE lineages to the cortical interneuron population is higher than previously estimated (about 40% of the total). Of the cells fate mapped from the CGE, a large percentage are late-spiking interneurons expressing Reelin or bipolar VIP/calretinin RSNP populations (G. Miyoshi *et al.*, unpublished results).

3.6. A question of semantics? Current definition of the eminences

As previously discussed, historically, the division of the ganglionic eminences has been based on anatomical landmarks, which are subjective, change over time, and may or may not reflect the local molecular identity. The lack of genetic markers that clearly differentiate the lineages that arise from the distinct progenitor pools within the eminences makes the division and definition of the ventral progenitor territories complex. Of these the best understood is the MGE, which gives rise to both SST and PV interneuron populations. While proportionally, the PV interneurons appear to be derived from the ventral portion of the MGE, and SST interneurons arise from the dorsal MGE (Flames *et al.*, 2007), whether these are derived from a common progenitor pool or not is unclear. The distinction between the LGE and CGE is more ambiguous. This has led some scientists to integrate the CGE within the LGE domain (Flames *et al.*, 2007). However, the CGE and LGE clearly give rise to distinct neuronal populations, despite the fact that certain neuronal subtypes are produced in both regions (Butt *et al.*, 2005; Xu *et al.*, 2004).

3.7. MGE- and CGE-derived cortical interneurons are generated with different temporal profiles

It has long been recognized that, as a whole, cortical interneurons show the same inside–out pattern of generation that is observed in pyramidal cells (Anderson *et al.*, 2002; Cavanagh and Parnavelas, 1989). Different subtypes of MGE-derived interneurons are produced at particular developmental times. Early-born MGE-derived interneurons end up in deep layers, while late-born interneurons occupy superficial layers of the mature cortex, and each cohort possesses unique electrophysiological properties characteristic of their birthdate (Miyoshi *et al.*, 2007). Moreover, heterochronic transplants of progenitors demonstrated that the fate of transplanted cells is determined by the age of the donor and not the age of the recipient (Butt *et al.*, 2005). Both pieces of evidence suggest that MGE progenitors have different potentials at different times (see figure 3.3).

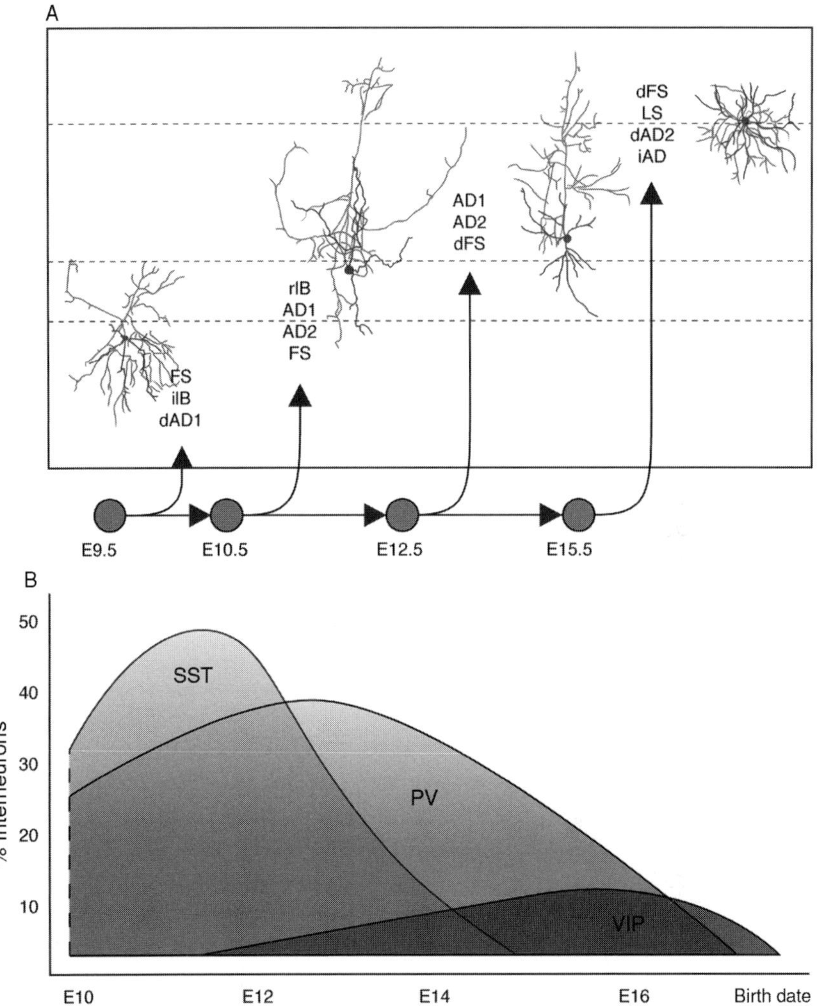

Figure 3.3 Temporal origin of cortical interneurons. (A) Examples of the morphological and electrophysiological diversity of cortical interneurons derived from the MGE at different times. Axons and dendrites are indicated in red and blue, respectively. This figure is adapted from Miyoshi *et al.* (2007). (B) Diagram of temporal origin of three subtypes of MGE interneurons, the somatostatin (SST—pink), the parvalbumin (PV—orange), and a subtype of CGE-derived interneuron (VIP—blue). (See Color Insert.)

While most of MGE-derived interneurons are born around E13.5, the generation of CGE-derived interneurons peaks at later ages (around E15.5) (Butt *et al.*, 2005; Nery *et al.*, 2002). Preliminary transplant results suggested that the interneurons generated in the CGE also differ based on birthdate (Butt *et al.*, 2005). However, more accurate genetic fate-mapping analysis have recently shown that, contrary to the MGE, the subtypes of interneurons generated within the CGE does not significantly change over time. Furthermore, the birthdate of CGE neurons does not predict their cortical layer destination, unlike their MGE counterparts (G. Miyoshi *et al.*, unpublished results). In fact, at least one subpopulations of CGE-derived cells, the CR interneurons, have an outside–in gradient of generation (Rymar and Sadikot, 2007), while VIP interneuron layer specificity does not appear to be correlated with their time of birth (Cavanagh and Parnavelas, 1989).

3.8. Where and when is interneuron identity specified?

As discussed earlier, there are now multiple lines of evidence showing that an interneuron's subtype identity is related to both its temporal and spatial embryonic origin. This correlation likely reflects the logic by which the genetic program functioning in progenitors and migrating interneurons is initiated. But, how is interneuron diversity *per se* established? This question has been a central point of debate in the field over the last decade. Despite a multitude of attempts to reveal the mechanisms by which an interneuron acquires its mature properties, we still do not have a definitive answer to this question. However, based on the insights discussed above, it is perhaps not premature to begin to hypothesize on how this process might be regulated.

One school of thought (championed by the Rubenstein laboratory and colleagues) is that, akin to what happens in the spinal cord, the ganglionic eminences contain multiple distinct pools of progenitors, each of which gives rise to specific interneuron subtypes. In the spinal cord, analogous pools are defined by their expression of a specific combination of transcription factors established by a Shh morphogen gradient (Jessell, 2000). Even though the subpallial eminences are responsive to Shh gradients, and the MGE and CGE broadly give rise to different groups of interneuron subtypes, the identification of progenitor pools that are restricted to producing particular interneuron subtypes has yet to be definitively demonstrated. Using a combinatorial approach, based on the examination of the patterns and levels of expression of different transcription factors, Flames *et al.* (2007) divided both the MGE and LGE eminences into nine subdomains. However, this effort to map the eminences into microdomains is too coarse to account for the diversity of interneuron subtypes that arise from the MGE (Miyoshi *et al.*, 2007). To test the specificity of these microdomains more directly, a microtransplantation fate-mapping study of two of the areas within the MGE was performed. Even though there was a bias in the interneuron subtypes arising from each of the two areas transplanted, each MGE subregion still produced a mix of interneuron subtypes. Therefore, although the hypothesis of distinct MGE progenitor subdomains specified through the actions of unique transcriptional combinations is very attractive, it has yet to be established.

An alternative hypothesis is that, similar to what happens in the retina (Cepko *et al.*, 1996), and the pyramidal cells of the cerebral cortex (Desai and McConnell, 2000; Rakic, 1974), interneuron diversity depends on changes in progenitor competence over time. In this model, progenitors within the same region go through systematic alterations in their potential as development progresses. This hypothesis is in concordance with the different electrophysiological characteristics of the MGE-derived interneurons produced at distinct developmental time points (Miyoshi *et al.*, 2007). However, this model cannot completely account for the generation of interneuron diversity, both because the correlation between cell type and birthdate is only approximate, and because the MGE may account for the generation of as little as 60% of all cortical interneurons (Fig. 3.3).

3.9. The contribution of space and time: Lessons from the Nkx2.1 progenitors

The question of mechanism aside, the above studies provide clear evidence in favor of the idea that interneuron subtype identity is assigned, at least partially, at the progenitor level. How might a blend of these hypotheses account for the coordinated regulation of interneuron subtype identity?

In favor of transcription factors regulating spatial identity, *Nkx2.1* appears to positively regulate MGE fates, while repressing LGE/CGE fates (Butt *et al.*, 2008; Sussel *et al.*, 1999). This suggests that *Nkx2.1* might be acting as a molecular toggle switch similar to how homologous genes function in the spinal cord, where type 1 genes (repressed by Shh) and type 2 genes (induced by Shh) compete to determine cell fate. Following this idea, competitive crossrepression in the telencephalon may lead to the coalescence of distinct progenitor domains that determine specific neuronal subtypes (Jessell, 2000).

Nkx2.1 per se is unlikely to regulate changes in the temporal competence of progenitors, as its expression is relatively constant through development. Hence if the progenitors within the MGE change their potential over time, *Nkx2.1* expression while being necessary cannot be sufficient. The genetic cascades regulating the temporal competence of Drosophila nerve cord progenitor cells have been shown to be regulated by the sequential expression of Hunchback, Kruppel, Pdm, and Castor, which collectively result in the ordered production of different neuronal subtypes from a common lineage (Isshiki *et al.*, 2001). Intriguingly, homologs of many of these genes are expressed within the telencephalon but not with patterns that suggest that this genetic network is precisely mirrored in mammalian brain (Jacob *et al.*, 2008). However, hints of analogous temporally regulated hierarchies within the telencephalon have emerged, particularly in the pallium. For instance, the production of the earliest pallial populations, the Cajal Retzius cells, appear to be negatively regulated by *FoxG1* (Hanashima *et al.*, 2004; Shen *et al.*, 2006). Moreover, an intriguing interplay between *Fezf2*, *Tbr2*, *Sox5*, and *Satb2* appears to control the sequential production of deep and superficial projection neurons in the cortex (Alcamo *et al.*, 2008; Britanova *et al.*, 2008; Chen *et al.*, 2008; Hevner, 2006; Lai *et al.*, 2008; Molyneaux *et al.*, 2005). As yet, complementary networks in progenitor cells that give rise to cortical interneurons have not been identified, but likely exist. As they emerge, the logic by which spatial identity genes such as *Nkx2.1* are coregulated with temporal identity genes should be forthcoming.

One possibility is that spatial distinctions between the MGE and CGE are in fact better thought of as sequential temporal phases that sweep over these eminences as development progresses. In this view, the MGE and CGE change their competence as a result of the changing expression of genes such as *Nkx2.1*. Viewed in this manner, the MGE extinguishes expression of *Nkx2.1* during later development, allowing cortical interneuron subtypes that we consider to be CGE-derived to be generated. Whether such an extreme view obviates the idea that cortical interneuron subtypes are at least partially depending on spatially discrete progenitor zones seems unlikely. It does, however, emphasize the point that the differences between spatial and temporal mechanisms may be somewhat semantic. Finally, in considering the intrinsic mechanisms by which cortical interneuron subtype is determined, there is the question of lineage. Specifically we must determine whether there are stereotypic clonal relationships between particular cortical interneuron subtypes that arise from a common progenitor pool. To address this question, it will be necessary to do a lineage analysis similar to what has been done previously in the retina and cortex (Cepko *et al.*, 1996).

3.10. The influence of postmitotic environment on cortical interneuron diversity

So far, we only have addressed the mechanisms of generating interneuron diversity that result from intrinsic signaling in the progenitor pool that gives rise to these neurons. Certainly, existing data support the idea that the specification of interneuron subtypes is dependent on events occurring prior to these lineages becoming postmitotic. For example, it has been shown that heterotopic transplants of cortical interneurons at E13.5 does not alter their subtype identity, in that MGE cells transplanted in the CGE still adopt an MGE fate and CGE cells transplanted into the MGE still adopt an CGE fate (Butt *et al.*, 2005). However, even if identity is first specified early, it is only determined postnatally, and therefore several checkpoints must exist. Alternatively, the ability of transplanted cells to maintain their normal fate may be simply a product of their ability to migrate to their appropriate destination from an ectopic location. In this model, their ability to follow the correct postmitotic cues through directed migration is the key to attaining their proper subtype identity. At least one study indicates that when their normal migratory route is disrupted, so is their laminar position.

Heterochronic injections of E12.5 and E15.5 MGE cells in the telencephalic ventricles found that these transplanted cells adopted a laminar fate dependent on the age of the host rather than on their intrinsic birthdate (Valcanis and Tan, 2003). This study, however, did not explore the intrinsic properties of these transplanted interneurons and therefore it is not known whether challenging these cells in this manner altered their intrinsic subtype properties as well as their laminar distribution.

The question of the relative contribution of cell intrinsic versus nonintrinsic fate determinants has been more thoroughly explored in cortical pyramidal cells. Two opposing theories have been posited to account for pyramidal cell specification and arealization. One suggests that the mature characteristics of cortical subtypes are strongly influenced by their neocortical environment (McConnell, 1988; McConnell and Kaznowski, 1991; O'Leary and Stanfield, 1989). The alternative point of view was proposed in the "protomap hypothesis" (Rakic, 1988), in which the identity of the pyramidal cells is determined at the progenitor level. This concept implies that combinatorial gene expression during neurogenesis (Jessell, 2000) initiates an intrinsic program that ultimately results in the emergence of specific neuronal subtypes. Within the neocortex, evidence for the latter point of view came from landmark studies where the fate of pyramidal cells was challenged through heterotopic transplantation (McConnell, 1988; McConnell and Kaznowski, 1991) and from genetic studies where areal identity is maintained in the absence of thalamic afferents (Huffman *et al.*, 2004; Nakagawa *et al.*, 1999). Recently, genetic studies have uncovered some of the molecular effectors that mediate these events such as specific transcription factors that control pyramidal cell fate and areal identity (Chen *et al.*, 2005a,b; Cholfin and Rubenstein, 2007; Hamasaki *et al.*, 2004; Molyneaux *et al.*, 2005). It is our view that even though many of the components of cell identity, such as general intrinsic physiological properties and layer position, are likely determined at the progenitor level, other components such as choice of synaptic partners are likely influenced by local environmental cues. Therefore, these two mechanisms may contribute to different aspects of cell specification. While these two models have been classically used to describe pyramidal cells, we think that cortical interneurons should be integrated within the logic of this existing framework, and are likely subject to similar rules.

The extent to which the specific final properties of interneurons are determined at a progenitor level or acquired them later is still unclear. Taking into account the different models posited and the studies to date, we envision interneuron fate specification as a cladistic process (an idea raised in discussion between Gord Fishell and Kenneth Harris). By this we mean that the potential of progenitors narrows over time through a set of sequential restrictions that are initiated during the progenitor stage but persist postmitotically. While early decisions, such as the decision to be a PV or a SST interneuron are likely established at the progenitor level, later restrictions occur over time as they migrate, integrate into the cortex, and establish the first contacts with other cells (see figure 3.4). We think that this progression is mediated by a combination of genes expressed by the cells throughout their life. Therefore, genetic profiling of these cells at critical developmental stages will bring about new insights into interneuron development and subtype specification. Recently, we showed that migrating embryonic interneuron precursors express a variety of transcription factors and mature interneuron subtype markers, supporting the idea that significant aspects of their identity are already acquired during embryonic stages (Batista-Brito *et al.*, 2008) (Fig. 3.4).

3.11. Modes of migration of cortical interneurons—The high road and the low road

Cortical interneurons originating in the subpallium migrate tangentially to enter the cortex (Corbin *et al.*, 2001; Marin and Rubenstein, 2001). Tangentially migrating interneurons enter the cortical plate through two different migratory routes: a superficial path within the marginal zone (MZ); and a deep route positioned at the subplate/SVZ interface (Lavdas *et al.*, 1999). The molecular cues guiding interneurons to each of these routes are unknown, as is whether there exist preferred routes for different interneuron subtypes.

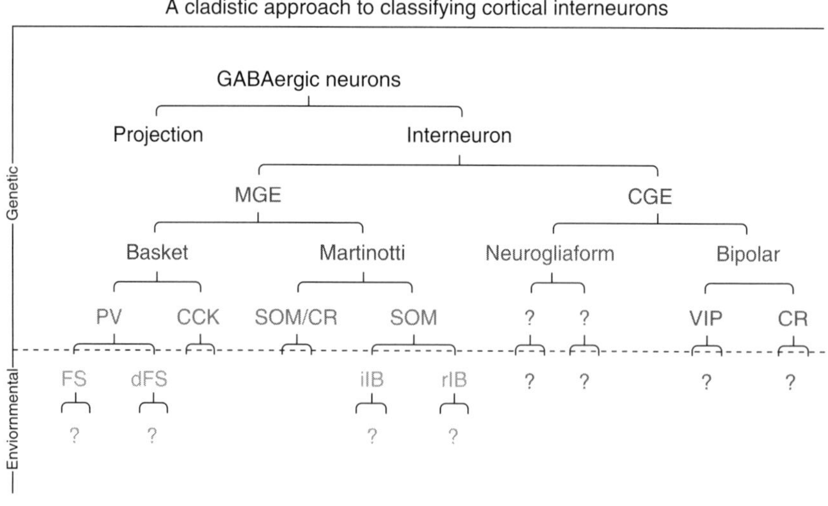

Figure 3.4 Cladistic hypothesis of cortical interneuron specification. We believe the fate of inhibitory neuron subtypes are sequentially influenced by both intrinsic genetic cues within the progenitor populations and environmental signals experienced by these populations postmitotically. In this model, we suggest that while the cardinal subdivisions are dependent on an intrinsic (genetic) program, the secondary subdivisions are determined later by environmental signals.

Tangential migration is mediated by a combination of molecular cues that function to both selectively repel and attract cortical interneuron populations (Marin and Rubenstein, 2003). While this process is directly influenced by the guidance molecules themselves (e.g., Semaphorins, Slits, neuregulin), the responsiveness of migrating interneurons to these cues is also regulated by subtype specific transcription factors. Recently, a wonderful example of this comes from work by the Marin group, who demonstrated that the semaphorin receptor Nrp2 is negatively regulated by *Nkx2.1* (Nobrega-Pereira *et al.*, 2008). As a result the Sema3a (the repellent ligand for the Nrp2 receptor) expression in the striatum prevents cortical interneurons from invading this structure. By contrast, the persistent Nkx2.1 expression in interneurons destined for the striatum prevents Nrp2 expression in this population and as a result permits them to invade this region. Transcription factors have also been shown to control the intracellular machinery required for migration. *Dlx1/2* compound double mutants have impaired cortical interneuron migration (Anderson *et al.*, 1997b) at least in part through their control of the p21-activated serine/threonine kinase PAK3, a downstream effector of the Rho family of GTPases (Cobos *et al.*, 2007). Finally, ambient GABA also influences interneuron migration (Cuzon *et al.*, 2006; Lopez-Bendito *et al.*, 2003; Manent and Represa, 2007; Manent *et al.*, 2005; Poluch and Juliano, 2007).

After a period of tangential migration, interneurons shift to a radial mode of migration when they enter the cortical plate (Ang *et al.*, 2003; Polleux *et al.*, 2002). Little is known about the mechanisms behind this switch in their mode of migration; however, the timing of this shift seems to be regulated by chemokines. Specifically, Cxcl12, a chemokine expressed by both the meninges and pyramidal cells, and Cxcr4, Cxcl12 receptors expressed in interneurons (Stumm *et al.*, 2003; Tiveron *et al.*, 2006) appear to jointly mediate this process. The impairment of this signaling pathway due to the either the absence of the chemokine (Cxcl12) or its receptor (Cxcr4) leads to the premature invasion of the cortical plate by migrating interneurons, and subsequent abnormal interneuron lamination (Lopez-Bendito *et al.*, 2008; Tiveron *et al.*, 2006). The exact mechanisms directing interneurons to stop migrating when they reach their final position are still poorly understood.

4. The Integration of Interneurons into Cortical Networks

To this point, we have examined the diversity of cortical interneurons based on their time of origin, or where within the ventral telencephalon they arise. However, at birth, the emergence of specific cortical interneuron subclasses is far from complete. In this section, we will ask how is complexity assembled, by examining emergent interneuron properties and network integration during postnatal development.

A complex set of developmental steps are involved in the integration of interneurons into the cortical network. These require the interplay of intrinsic genetic programs and their modulation by cell–cell and matrix–cell interactions. Some of the more notable of these are the development of reciprocal connections between interneurons and pyramidal cells, the transition of GABAergic signaling from excitation to inhibition and the maturation of synaptic connectivity through consolidation and refinement. Each of these contributes significantly to the establishment of a mature functional network during the first few postnatally weeks.

4.1. GABA is excitatory during development

In adult mammals, GABA is the main inhibitory neurotransmitter. GABA acts through ionotropic receptor channels, which are permeable to anions (namely Cl^- and bicarbonate). In mature neurons, activation of $GABA_A$-R leads to chloride influx, due to low intracellular chloride levels. GABA receptor activation therefore generally results in neurons becoming more hyperpolarized. For a long time, it has been known that GABA in some contexts is excitatory. The excitatory actions of GABA were documented on cultured neurons of embryonic chick spinal cord three decades ago (Obata *et al.*, 1978). In 1984, it was shown that GABA application leads to depolarization of neonatal hippocampal pyramidal neurons (Mueller *et al.*, 1984); however, the mechanism by which this process occurred was only discovered 5 years later. In a landmark study, Ben-Ari and colleagues showed that developing networks are spontaneously active due to the excitatory actions of GABAergic transmission in early development. This work demonstrated that at the end of the first postnatal week, GABA signaling onto hippocampal CA1 neurons resulted in hyperpolarization due to a developmental decrease of intracellular concentration of Cl^- over time (leading to a switch in the Cl^- reversal potential) (Ben-Ari *et al.*, 1989). This discovery was further confirmed in multiple species demonstrating that this shift represents an evolutionarily conserved mechanism. It was also observed in neurons in other regions of the brain (Ben-Ari *et al.*, 2007) including the cortex (Gulledge and Stuart, 2003; LoTurco *et al.*, 1995; Luhmann and Prince, 1991; Martina *et al.*, 2001; Owens *et al.*, 1996), as well as in peripheral structures.

Intracellular chloride concentration is actively regulated by two transporters NKCC1 and KCC2. NKCC1 imports Cl^- into the cell and is highly expressed in immature neurons. Conversely, KCC2, a Cl^- exporter, expression increases after the first postnatal week (Ben-Ari, 2002; Dzhala *et al.*, 2005). The decreased concentration of intracellular Cl^- as neurons mature has been attributed to the postnatal onset of expression of the Cl^-

exporter KCC2. That the evolutionary alteration in the expression of a single transporter protein could have such a marked effect on GABAergic signaling is remarkable.

4.2. Early activity patterns

The immature cerebral cortex self-organizes into local neuronal clusters long before it is stimulated by sensory input (Katz and Crowley, 2002). Spontaneous neuronal activity is a characteristic of developing networks and is essential for their maturation (Cang *et al.*, 2005; Kandler and Gillespie, 2005; Katz and Shatz, 1996; Khazipov *et al.*, 2004; Nicol *et al.*, 2007). In the neocortex, synchronous spontaneous calcium activity starts early during development, and in fact is already initiated in progenitors (Owens and Kriegstein, 1998). In postnatal animals, cortical neurons still display coherent spontaneously activity (Garaschuk *et al.*, 2000; Khazipov *et al.*, 2004; Yuste *et al.*, 1992). In addition, they begin to communicate through both transmitter-gated receptors (Flint *et al.*, 1999; Peinado, 2000), as well as electrical and chemical synapses. In the newborn mouse, the immature cortical network initiates calcium activity within cortical columns, which is at first synchronized by gap junctions (Dupont *et al.*, 2006; Kandler and Katz, 1998), and then mediated by a synaptic network dependent on NMDA (*N*-methyl-d-aspartate) receptors (Dupont *et al.*, 2006). In addition, in both the hippocampus and neocortex, synaptically mediated activity is preceded by intrinsic nonsynaptic calcium plateaus associated with membrane potential oscillations (Crepel *et al.*, 2007). These local and synchronous activity patterns named synchronous plateau assemblies (SPAs) are also generated by neurons coupled through gap junctions (Crepel *et al.*, 2007). SPAs are initiated around birth and are enhanced by oxytocin. SPA events begin to decline approximately coincident with the emergence of synaptic-driven network activity (Allène *et al.*, 2008) (Fig. 3.5).

So far, two different types of synapse-driven network patterns have been described, namely giant depolarizing potentials (GDPs) and cortical early network oscillations (cENOs). GDPs were the first developmental synaptic-driven cortical network oscillation discovered (Ben-Ari *et al.*, 1989). They emerge a few days after birth and are the first synapse-driven network pattern in the developing hippocampus (Ben-Ari *et al.*, 1989; Crepel *et al.*, 2007; Garaschuk *et al.*, 2000). GDPs are mediated by GABAergic synapses and ceases when GABA switches from excitatory to inhibitory (Ben-Ari *et al.*, 1989; Garaschuk *et al.*, 1998; Tyzio *et al.*, 2007). Until recently, the only counterparts of GDPs described for the cortex were cENOs. These are spontaneous Ca^{2+} waves that are observed within the cortex immediately after birth (Garaschuk *et al.*, 2000). Contrary to GDPs, cENOs are mediated by glutamatergic synapses and required activation of AMPA and NMDA receptors, but not $GABA_A$ receptors (Corlew *et al.*,

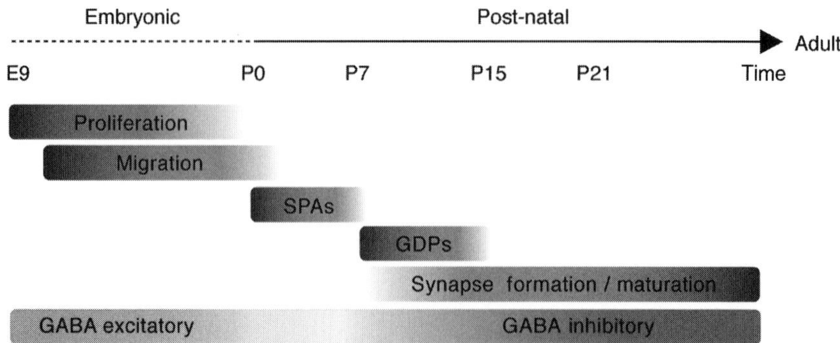

Figure 3.5 Summary of the essential events and processes that influence interneuron development. Embryonic ages are dominated by proliferation and cell migration. During this period GABA is excitatory. The first postnatal weeks are marked by early activity patterns, including synchronous plateau assemblies (SPAs) that are preceded by giant depolarizing potentials (GDPs). During this later period GABA switches from being excitatory to be inhibitory.

2004; Garaschuk *et al.*, 2000; McCabe *et al.*, 2006). Based on the presence of cENOs, it has been suggested that glutamatergic synapses, rather than GABAergic ones, have a privileged role on setting up early networks in the cortex (McCabe *et al.*, 2007). However, similarly to the hippocampus, in the developing cortex GABA is excitatory (Ben-Ari *et al.*, 2007) and recently it has been shown that in fact the neocortex also produces GABA-driven GDPs (Allène *et al.*, 2008). This form of cortical network activity, as in the hippocampus, occurs slightly later than cENOs (P6–P8 vs P0–P3 in rat) (Allène *et al.*, 2008) (see figure 3.5). Thus, GABAergic transmission and hence the interneurons contribute to spontaneous network oscillations in the developing cortex. Interneurons are therefore very likely to play an important role in postnatal network construction through theses synapse-driven coordinated activity patterns. In fact, it has been recently shown that spontaneous activity regulates GABA synthesis, affecting inhibitory innervation patterns (Chattopadhyaya *et al.*, 2007).

Major strides have been made over the last two decades. Even so the neuronal network and the cellular mechanisms underlying the process of cortical self-organization are still far from being comprehensively understood. Identification of other forms of nascent network activity and the interactions between these forms will be crucial for understanding how the establishment of cortical networks is achieved.

4.3. Interneuron development and neurological disorders

As we discussed previously, GABAergic interneurons have a powerful role in establishing networks and controlling their function. Abnormal development of GABAergic circuits and even a subtle imbalance in the ratio of excitatory versus inhibitory levels within the cerebral cortex may underlie many neurological and neurodevelopmental disorders such as epilepsy, autism spectrum disorders, schizophrenia, and Tourette's syndrome (Baraban and Tallent, 2004; Belmonte *et al.*, 2004; Cossart *et al.*, 2005; Dani *et al.*, 2005; Kalanithi *et al.*, 2005; Levitt *et al.*, 2004; Lewis *et al.*, 2005; Woo and Lu, 2006). With the recent implementation of high-throughput sequence techniques, there is currently a large effort to map genes associated with neurological disorders. By crosscorrelating these studies with the genes expressed in developing cortical interneurons, we recently came to the realization that many genes linked to autism, mental retardation, epilepsy, and schizophrenia are selectively expressed in cortical interneurons precursors (Batista-Brito *et al.*, 2008b). Consistent with this idea, a common observation associated with autistic spectrum disorders is the manifestation of "sharp spike" EEGs, suggestive of noisy and unstable cortical networks (Lewine *et al.*, 1999; Wheless and Kim, 2002).

It is our belief that the misspecification and/or dysfunction of specific subtypes of interneurons account for a significant portion of the etiology of many neurological disorders. As a consequence, we suggest this results in the improper consolidation of activity-based brain patterning during early postnatal development. Hence, we hypothesize that abnormal development of cortical interneuron subtypes during late embryogenesis due to environment perturbations coupled with genetic abnormalities might represent a primary cause for many neurodevelopment disorders. Consistent with this idea, the postconception days 20–24 in humans (equivalent to E11.5–E14.5 in mice) appears to be a susceptibility period for environmental perturbations, such as thalidomide or valproic acid. Exposure to these factors significantly increases the risk of cerebral dysfunction in both humans and rodents (Arndt *et al.*, 2005; Miller *et al.*, 2005). Notably, this is the precise developmental stage during which cortical interneuron diversity is being established (Butt *et al.*, 2005; Wonders and Anderson, 2006). Moreover, perturbation of certain cortical interneuronal subtypes (double bouquet cells and parvalbumin-positive cells) have been reported in patients afflicted with autism (Casanova *et al.*, 2002). Additionally, *GABARB3* and *Arx*, genes central to cortical interneuronal function, are within susceptibility loci associated with autism spectrum disorders (Chaste *et al.*, 2007; Vincent *et al.*, 2006). Similarly, *Npas1* and *Npas3* are both selectively expressed in migrating and mature cortical interneurons and have been implicated in schizophrenia (Batista-Brito *et al.*, 2008b; Pieper *et al.*, 2005). Another example of how developmentally expressed transcription factors may provide clues for understanding neurodevelopmental disorders comes from genetic studies of Rett's syndrome. This syndrome leads to mental retardation and is caused by X-linked mutations in MeCP2 (Moretti and Zoghbi, 2006). It has been shown that *Dlx5*, a gene essential for interneuron development (Stuhmer *et al.*, 2002), is a direct target of MeCP2 (Horike *et al.*, 2005). Finally, *Nrg1*, expressed in migrating cortical interneurons (Flames *et al.*, 2004), is located in a susceptibility locus for schizophrenia (Corfas *et al.*, 2004).

In summary, due to the complexity and heterogeneity of GABAergic circuits, the mechanisms by which they develop and function have been difficult to study. However, with the burgeoning of sophisticated new approaches, these issues are rapidly becoming tractable. Techniques, such as the use of developmental genetics for specific interneuron subtype labeling and high-resolution imaging, are contributing toward our understanding of the principles underlying the establishment of GABAergic networks. Although the association between neurological disorders and interneuron function is already substantial, our understanding of this connection is in its nascence. Analyzing the functional and behavioral consequences of targeted modifications in these genes holds the promise of providing mouse models for the study of various neurodevelopmental disorders. It seems likely that the intrinsic programs that drive cortical interneuron development are causally linked to the integration and maintenance of these cells in cortical circuitry. This in turn will no doubt help elucidate the functions of these genes in both normal and abnormal brain function.

ACKNOWLEDGMENTS

We would like to thank Rosa Cossart, Bernardo Rudy, Robert Machold, Elsa Rossignol, Jens Herling-Leffler, Theofanis Karayannis, and Edmund Au for critically revising this manuscript. We would like to thank Karl-Johan Hjerling for his great graphic support on creating the figures of this manuscript.

REFERENCES

Aboitiz, F., and Montiel, J. (2007). Co-option of signaling mechanisms from neural induction to telencephalic patterning. *Rev. Neurosci.* **18,** 311–342.

Alcamo, E. A., Chirivella, L., Dautzenberg, M., Dobreva, G., Farinas, I., Grosschedl, R., and McConnell, S. K. (2008). Satb2 regulates callosal projection neuron identity in the developing cerebral cortex. *Neuron* **57,** 364–377.

Allène, C., Cattani, A., Ackman, J. B., Bonifazi, P., Aniksztejn, L., Ben-Ari, Y., and Cossart, R. (2008). Sequential generation of two distinct synapse-driven network patterns in developing neocortex. *J. Neurosci.* **28,** 12851–12863.

Anderson, S. A., Eisenstat, D. D., Shi, L., and Rubenstein, J. L. (1997a). Interneuron migration from basal forebrain to neocortex: Dependence on Dlx genes [see comment]. *Science* **278,** 474–476.

Anderson, S. A., Kaznowski, C. E., Horn, C., Rubenstein, J. L., and McConnell, S. K. (2002). Distinct origins of neocortical projection neurons and interneurons *in vivo*. *Cereb. Corte.* **12,** 702–709.

Anderson, S. A., Marin, O., Horn, C., Jennings, K., and Rubenstein, J. L. (2001). Distinct cortical migrations from the medial and lateral ganglionic eminences. *Development* **128,** 353–363.

Anderson, S. A., Qiu, M., Bulfone, A., Eisenstat, D. D., Meneses, J., Pedersen, R., and Rubenstein, J. L. (1997b). Mutations of the homeobox genes Dlx-1 and Dlx-2 disrupt the striatal subventricular zone and differentiation of late born striatal neurons. *Neuron* **19,** 27–37.

Ang, E. S., Jr., Haydar, T. F., Gluncic, V., and Rakic, P. (2003). Four-dimensional migratory coordinates of GABAergic interneurons in the developing mouse cortex. *J. Neurosci.* **23,** 5805–5815.

Arndt, T. L., Stodgell, C. J., and Rodier, P. M. (2005). The teratology of autism. *Int. J. Dev. Neurosci.* **23,** 189–199.

Baraban, S. C., and Tallent, M. K. (2004). Interneuron Diversity series: Interneuronal neuropeptides-endogenous regulators of neuronal excitability. *Trends Neurosci.* **27,** 135–142.

Batista-Brito, R., Close, J., Machold, R., and Fishell, G. (2008a). The distinct temporal origins of olfactory bulb interneuron subtypes. *J. Neurosci.* **28,** 3966–3975.

Batista-Brito, R., Machold, R., Klein, C., and Fishell, G. (2008b). Gene expression in cortical interneuron precursors is prescient of their mature function. *Cereb. Corte.* **18,** 2306–2317.

Belmonte, M. K., Cook, E. H., Jr., Anderson, G. M., Rubenstein, J. L., Greenough, W. T., Beckel-Mitchener, A., Courchesne, E., Boulanger, L. M., Powell, S. B., Levitt, P. R., Perry, E. K., Jiang, Y. H., *et al.* (2004). Autism as a disorder of neural information processing: Directions for research and targets for therapy. *Mol. Psychiatry* **9,** 646–663.

Ben-Ari, Y. (2002). Excitatory actions of gaba during development: The nature of the nurture. *Nat. Rev. Neurosci.* **3,** 728–739.

Ben-Ari, Y., Cherubini, E., Corradetti, R., and Gaiarsa, J. L. (1989). Giant synaptic potentials in immature rat CA3 hippocampal neurones. *J. Physiol.* **416,** 303–325.

Ben-Ari, Y., Gaiarsa, J. L., Tyzio, R., and Khazipov, R. (2007). GABA: A pioneer transmitter that excites immature neurons and generates primitive oscillations. *Phys. Rev.* **87,** 1215–1284.

Britanova, O., de Juan Romero, C., Cheung, A., Kwan, K. Y., Schwark, M., Gyorgy, A., Vogel, T., Akopov, S., Mitkovski, M., Agoston, D., Sestan, N., Molnár, Z., *et al.* (2008). Satb2 is a postmitotic determinant for upper-layer neuron specification in the neocortex. *Neuron* **57,** 378–392.

Bulfone, A., Puelles, L., Porteus, M. H., Frohman, M. A., Martin, G. R., and Rubenstein, J. L. (1993). Spatially restricted expression of Dlx-1, Dlx-2 (Tes-1), Gbx-2, and Wnt-3 in the embryonic day 12.5 mouse forebrain defines potential transverse and longitudinal segmental boundaries. *J. Neurosci.* **13,** 3155–3172.

Bulfone, A., Wang, F., Hevner, R., Anderson, S., Cutforth, T., Chen, S., Meneses, J., Pedersen, R., Axel, R., and Rubenstein, J. L. (1998). An olfactory sensory map develops in the absence of normal projection neurons or GABAergic interneurons. *Neuron* **21,** 1273–1282.

Butt, S. J., Fuccillo, M., Nery, S., Noctor, S., Kriegstein, A., Corbin, J. G., and Fishell, G. (2005). The temporal and spatial origins of cortical interneurons predict their physiological subtype [see comment]. *Neuron* **48,** 591–604.

Butt, S. J., Sousa, V. H., Fuccillo, M. V., Hjerling-Leffler, J., Miyoshi, G., Kimura, S., and Fishell, G. (2008). The requirement of Nkx2–1 in the temporal specification of cortical interneuron subtypes [see comment]. *Neuron* **59,** 722–732.

Cang, J., Renteria, R. C., Kaneko, M., Liu, X., Copenhagen, D. R., and Stryker, M. P. (2005). Development of precise maps in visual cortex requires patterned spontaneous activity in the retina. *Neuron* **48,** 797–809.

Casanova, M. F., Buxhoeveden, D. P., Switala, A. E., and Roy, E. (2002). Minicolumnar pathology in autism. *Neurology* **58,** 428–432.

Casarosa, S., Fode, C., and Guillemot, F. (1999). Mash1 regulates neurogenesis in the ventral telencephalon. *Development* **126,** 525–534.
Cauli, B., Audinat, E., Lambolez, B., Angulo, M. C., Ropert, N., Tsuzuki, K., Hestrin, S., and Rossier, J. (1997). Molecular and physiological diversity of cortical nonpyramidal cells. *J. Neurosci.* **17,** 3894–3906.
Cavanagh, M. E., and Parnavelas, J. G. (1989). Development of vasoactive-intestinal-polypeptide-immunoreactive neurons in the rat occipital cortex: A combined immuno-histochemical-autoradiographic study. *J. Comp. Neurol.* **284,** 637–645.
Cepko, C. L., Austin, C. P., Yang, X., Alexiades, M., and Ezzeddine, D. (1996). Cell. fate determination in the vertebrate retina. *Proc. Natl. Acad. Sci. USA* **93,** 589–595.
Chaste, P., Nygren, G., Anckarsater, H., Rastam, M., Coleman, M., Leboyer, M., Gillberg, C., and Betancur, C. (2007). Mutation screening of the ARX gene in patients with autism. *American Journal of Medical Genetics. Part B, Neuropsychiatric Genetics: The Official Publication of the International Society of Psychiatric Genetics* **144,** 228–230.
Chattopadhyaya, B., Di Cristo, G., Wu, C. Z., Knott, G., Kuhlman, S., Fu, Y., Palmiter, R. D., and Huang, Z. J. (2007). GAD67-mediated GABA synthesis and signaling regulate inhibitory synaptic innervation in the visual cortex. *Neuron* **54,** 889–903.
Chen, B., Schaevitz, L. R., and McConnell, S. K. (2005a). Fezl regulates the differentiation and axon targeting of layer 5 subcortical projection neurons in cerebral cortex. *Proc. Natl. Acad. Sci. USA* **102,** 17184–17189.
Chen, B., Wang, S. S., Hattox, A. M., Rayburn, H., Nelson, S. B., and McConnell, S. K. (2008). The Fezf2-Ctip2 genetic pathway regulates the fate choice of subcortical projection neurons in the developing cerebral cortex. *Proc. Natl. Acad. Sci. USA* **105,** 11382–11387.
Chen, J. G., Rasin, M. R., Kwan, K. Y., and Sestan, N. (2005b). Zfp312 is required for subcortical axonal projections and dendritic morphology of deep-layer pyramidal neurons of the cerebral cortex. *Proc. Natl. Acad. Sci. USA* **102,** 17792–17797.
Cholfin, J. A., and Rubenstein, J. L. (2007). Patterning of frontal cortex subdivisions by Fgf17. *Proc. Natl. Acad. Sci. USA* **104,** 7652–7657.
Chow, A., Erisir, A., Farb, C., Nadal, M. S., Ozaita, A., Lau, D., Welker, E., and Rudy, B. (1999). K(+) channel expression distinguishes subpopulations of parvalbumin- and somatostatin-containing neocortical interneurons. *J. Neurosci.* **19,** 9332–9345.
Cobb, S. R., Buhl, E. H., Halasy, K., Paulsen, O., and Somogyi, P. (1995). Synchronization of neuronal activity in hippocampus by individual GABAergic interneurons. *Nature* **378,** 75–78.
Cobos, I., Calcagnotto, M. E., Vilaythong, A. J., Thwin, M. T., Noebels, J. L., Baraban, S. C., and Rubenstein, J. L. (2005). Mice lacking Dlx1 show subtype-specific loss of interneurons, reduced inhibition and epilepsy [see comment]. *Nat. Neurosci.* **8,** 1059–1068.
Cobos, I., Puelles, L., and Martinez, S. (2001). The avian telencephalic subpallium originates inhibitory neurons that invade tangentially the pallium (dorsal ventricular ridge and cortical areas). *Dev. Bio.* **239,** 30–45.
Corbin, J. G., Gaiano, N., Machold, R. P., Langston, A., and Fishell, G. (2000). The Gsh2 homeodomain gene controls multiple aspects of telencephalic development. *Development* **127,** 5007–5020.
Corbin, J. G., Nery, S., and Fishell, G. (2001). Telencephalic cells take a tangent: Non-radial migration in the mammalian forebrain. *Nat. Neurosci.* **4,** 1177–1182.
Corfas, G., Roy, K., and Buxbaum, J. D. (2004). Neuregulin 1-erbB signaling and the molecular/cellular basis of schizophrenia. *Nat. Neurosci.* **7,** 575–580.
Corlew, R., Bosma, M. M., and Moody, W. J. (2004). Spontaneous, synchronous electrical activity in neonatal mouse cortical neurones. *J. Physiol.* **560,** 377–390.
Cossart, R., Bernard, C., and Ben-Ari, Y. (2005). Multiple facets of GABAergic neurons and synapses: Multiple fates of GABA signalling in epilepsies. *Trends Neurosci.* **28,** 108–115.

Crepel, V., Aronov, D., Jorquera, I., Represa, A., Ben-Ari, Y., and Cossart, R. (2007). A parturition-associated nonsynaptic coherent activity pattern in the developing hippocampus. *Neuron* **54,** 105–120.

Cuzon, V. C., Yeh, P. W., Cheng, Q., and Yeh, H. H. (2006). Ambient. GABA promotes cortical entry of tangentially migrating cells derived from the medial ganglionic eminence. *Cereb. Corte.* **16,** 1377–1388.

Dani, V. S., Chang, Q., Maffei, A., Turrigiano, G. G., Jaenisch, R., and Nelson, S. B. (2005). Reduced cortical activity due to a shift in the balance between excitation and inhibition in a mouse model of Rett syndrome. *Proc. Natl. Acad. Sci. USA* **102,** 12560–12565.

de Carlos, J. A., Lopez-Mascaraque, L., and Valverde, F. (1996). Dynamics of cell migration from the lateral ganglionic eminence in the rat. *J. Neurosci.* **16,** 6146–6156.

Deacon, T. W., Pakzaban, P., and Isacson, O. (1994). The lateral ganglionic eminence is the origin of cells committed to striatal phenotypes: Neural transplantation and developmental evidence. *Brain Research* **668,** 211–219.

DeFelipe, J. (1993). Neocortical neuronal diversity: Chemical heterogeneity revealed by colocalization studies of classic neurotransmitters, neuropeptides, calcium-binding proteins, and cell surface molecules. *Cereb. Corte.* **3,** 273–289.

Desai, A. R., and McConnell, S. K. (2000). Progressive restriction in fate potential by neural progenitors during cerebral cortical development. *Development* **127,** 2863–2872.

Du, T., Xu, Q., Ocbina, P. J., and Anderson, S. A. (2008). NKX2.1 specifies cortical interneuron fate by activating Lhx6. *Development* **135,** 1559–1567.

Dupont, E., Hanganu, I. L., Kilb, W., Hirsch, S., and Luhmann, H. J. (2006). Rapid developmental switch in the mechanisms driving early cortical columnar networks. *Nature* **439,** 79–83.

Dzhala, V. I., Talos, D. M., Sdrulla, D. A., Brumback, A. C., Mathews, G. C., Benke, T. A., Delpire, E., Jensen, F. E., and Staley, K. J. (2005). NKCC1 transporter facilitates seizures in the developing brain [see comment]. *Nat. Med.* **11,** 1205–1213.

Ferezou, I., Cauli, B., Hill, E. L., Rossier, J., Hamel, E., and Lambolez, B. (2002). 5-HT3 receptors mediate serotonergic fast synaptic excitation of neocortical vasoactive intestinal peptide/cholecystokinin interneurons. *J. Neurosci.* **22,** 7389–7397.

Fishell, G. (1997). Regionalization in the mammalian telencephalon. *Curr. Opin. Neurobiol.* **7,** 62–69.

Flames, N., Long, J. E., Garratt, A. N., Fischer, T. M., Gassmann, M., Birchmeier, C., Lai, C., Rubenstein, J. L., and Marin, O. (2004). Short- and long-range attraction of cortical GABAergic interneurons by neuregulin-1. *Neuron* **44,** 251–261.

Flames, N., Pla, R., Gelman, D. M., Rubenstein, J. L., Puelles, L., and Marin, O. (2007). Delineation of multiple subpallial progenitor domains by the combinatorial expression of transcriptional codes. *J. Neurosci.* **27,** 9682–9695.

Flint, A. C., Dammerman, R. S., and Kriegstein, A. R. (1999). Endogenous activation of metabotropic glutamate receptors in neocortical development causes neuronal calcium oscillations. *Proc. Natl. Acad. Sci. USA* **96,** 12144–12149.

Fogarty, M., Grist, M., Gelman, D., Marin, O., Pachnis, V., and Kessaris, N. (2007). Spatial genetic patterning of the embryonic neuroepithelium generates GABAergic interneuron diversity in the adult cortex. *J. Neurosci.* **27,** 10935–10946.

Freund, T. F. (2003). Interneuron Diversity series: Rhythm and mood in perisomatic inhibition. *Trends Neurosci.* **26,** 489–495.

Garaschuk, O., Hanse, E., and Konnerth, A. (1998). Developmental profile and synaptic origin of early network oscillations in the CA1 region of rat neonatal hippocampus. *J. Physiol.* **507,** 219–236.

Garaschuk, O., Linn, J., Eilers, J., and Konnerth, A. (2000). Large-scale oscillatory calcium waves in the immature cortex. *Nat. Neurosci.* **3,** 452–459.

Gerashchenko, D., Wisor, J. P., Burns, D., Reh, R. K., Shiromani, P. J., Sakurai, T., de la Iglesia, H. O., and Kilduff, T. S. (2008). Identification of a population of sleep-active cerebral cortex neurons. *Proc. Natl. Acad. Sci. USA* **105,** 10227–10232.

Goldberg, E. M., Clark, B. D., Zagha, E., Nahmani, M., Erisir, A., and Rudy, B. (2008). K+ channels at the axon initial segment dampen near-threshold excitability of neocortical fast-spiking GABAergic interneurons. *Neuron* **58,** 387–400.

Goldberg, E. M., Watanabe, S., Chang, S. Y., Joho, R. H., Huang, Z. J., Leonard, C. S., and Rudy, B. (2005). Specific functions of synaptically localized potassium channels in synaptic transmission at the neocortical GABAergic fast-spiking cell synapse. *J. Neurosci.* **25,** 5230–5235.

Gonchar, Y., and Burkhalter, A. (1997). Three distinct families of GABAergic neurons in rat visual cortex. *Cereb. Corte.* **7,** 347–358.

Gong, S., Zheng, C., Doughty, M. L., Losos, K., Didkovsky, N., Schambra, U. B., Nowak, N. J., Joyner, A., Leblanc, G., Hatten, M. E., and Heintz, N. (2003). A gene expression atlas of the central nervous system based on bacterial artificial chromosomes [see comment]. *Nature* **425,** 917–925.

Grigoriou, M., Tucker, A. S., Sharpe, P. T., and Pachnis, V. (1998). Expression and regulation of Lhx6 and Lhx7, a novel subfamily of LIM homeodomain encoding genes, suggests a role in mammalian head development. *Development* **125,** 2063–2074.

Gulledge, A. T., and Stuart, G. J. (2003). Excitatory actions of GABA in the cortex. *Neuron* **37,** 299–309.

Gupta, A., Wang, Y., and Markram, H. (2000). Organizing principles for a diversity of GABAergic interneurons and synapses in the neocortex [see comment]. *Science* **287,** 273–278.

Hamasaki, T., Leingartner, A., Ringstedt, T., and O'Leary, D. D. (2004). EMX2 regulates sizes and positioning of the primary sensory and motor areas in neocortex by direct specification of cortical progenitors [see comment]. *Neuron* **43,** 359–372.

Hanashima, C., Li, S. C., Shen, L., Lai, E., and Fishell, G. (2004). Foxg1 suppresses early cortical cell fate [see comment]. *Science* **303,** 56–59.

Hébert, J. M., and Fishell, G. (2008). The genetics of early telencephalon patterning: Some assembly required. *Nat. Rev. Neurosci.* [Epub ahead of print].

Hevner, R. F. (2006). From radial glia to pyramidal-projection neuron: Transcription factor cascades in cerebral cortex development. *Mol. Neurobiol.* **33,** 33–50.

Hirata, T., Li, P., Lanuza, G. M., Cocas, L.A., Huntsman, M. M., and Corbin, J. G. (2009). Identification of distinct telencephalic progenitor pools for neuronal diversity in the amygdala. *Nat. Neurosci.* **12,** 141–149.

Horike, S., Cai, S., Miyano, M., Cheng, J. F., and Kohwi-Shigematsu, T. (2005). Loss of silent-chromatin looping and impaired imprinting of DLX5 in Rett syndrome [see comment]. *Nat. Genet.* **37,** 31–40.

Huffman, K. J., Garel, S., and Rubenstein, J. L. (2004). Fgf8 regulates the development of intra-neocortical projections. *J. Neurosci.* **24,** 8917–8923.

Isshiki, T., Pearson, B., Holbrook, S., and Doe, C. Q. (2001). *Drosophila* neuroblasts sequentially express transcription factors which specify the temporal identity of their neuronal progeny. *Cell* **106,** 511–521.

Jacob, J., Maurange, C., and Gould, A. P. (2008). Temporal control of neuronal diversity: Common regulatory principles in insects and vertebrates? *Development* **135,** 3481–3489.

Jessell, T. M. (2000). Neuronal specification in the spinal cord: Inductive signals and transcriptional codes. *Nat. Rev. Genet.* **1,** 20–29.

Jonas, P., Bischofberger, J., Fricker, D., and Miles, R. (2004). Interneuron Diversity series: Fast in, fast out–temporal and spatial signal processing in hippocampal interneurons. *Trends Neurosci.* **27,** 30–40.

Kalanithi, P. S., Zheng, W., Kataoka, Y., DiFiglia, M., Grantz, H., Saper, C. B., Schwartz, M. L., Leckman, J. F., and Vaccarino, F. M. (2005). Altered parvalbumin-positive neuron distribution in basal ganglia of individuals with Tourette syndrome. *Proc. Natl. Acad. Sci. USA* **102,** 13307–13312.

Kamme, F., Salunga, R., Yu, J., Tran, D. T., Zhu, J., Luo, L., Bittner, A., Guo, H. Q., Miller, N., Wan, J., and Erlander, M. (2003). Single-cell microarray analysis in hippocampus CA1: Demonstration and validation of cellular heterogeneity. *J. Neurosci.* **23,** 3607–3615.

Kandler, K., and Gillespie, D. C. (2005). Developmental refinement of inhibitory sound-localization circuits. *Trends Neurosci.* **28,** 290–296.

Kandler, K., and Katz, L. C. (1998). Coordination of neuronal activity in developing visual cortex by gap junction-mediated biochemical communication. *J. Neurosci.* **18,** 1419–1427.

Katz, L. C., and Crowley, J. C. (2002). Development of cortical circuits: Lessons from ocular dominance columns. *Nat. Rev. Neurosci.* **3,** 34–42.

Katz, L. C., and Shatz, C. J. (1996). Synaptic activity and the construction of cortical circuits. *Science* **274,** 1133–1138.

Kawaguchi, Y., and Kubota, Y. (1996). Physiological and morphological identification of somatostatin- or vasoactive intestinal polypeptide-containing cells among GABAergic cell subtypes in rat frontal cortex. *J. Neurosci.* **16,** 2701–2715.

Kawaguchi, Y., and Kubota, Y. (1997). GABAergic cell subtypes and their synaptic connections in rat frontal cortex. *Cereb. Corte.* **7,** 476–486.

Khazipov, R., Sirota, A., Leinekugel, X., Holmes, G. L., Ben-Ari, Y., and Buzsaki, G. (2004). Early motor activity drives spindle bursts in the developing somatosensory cortex. *Nature* **432,** 758–761.

Kriegstein, A. R., and Noctor, S. C. (2004). Patterns of neuronal migration in the embryonic cortex. *Trends Neurosci.* **27,** 392–399.

Kubota, Y., and Kawaguchi, Y. (1994). Three classes of GABAergic interneurons in neocortex and neostriatum. *Japanese J. Physiol.* **44,** S145–S148.

Kubota, Y., and Kawaguchi, Y. (1997). Two distinct subgroups of cholecystokinin-immunoreactive cortical interneurons. *Brain Research* **752,** 175–183.

Lai, T., Jabaudon, D., Molyneaux, B. J., Azim, E., Arlotta, P., Menezes, J. R., and Macklis, J. D. (2008). SOX5 controls the sequential generation of distinct corticofugal neuron subtypes. *Neuron* **57,** 232–247.

Lau, D., Vega-Saenz de Miera, E. C., Contreras, D., Ozaita, A., Harvey, M., Chow, A., Noebels, J. L., Paylor, R., Morgan, J. I., Leonard, C. S., and Rudy, B. (2000). Impaired fast-spiking, suppressed cortical inhibition, and increased susceptibility to seizures in mice lacking Kv3.2 K+ channel proteins. *J. Neurosci.* **20,** 9071–9085.

Lavdas, A. A., Grigoriou, M., Pachnis, V., and Parnavelas, J. G. (1999). The medial ganglionic eminence gives rise to a population of early neurons in the developing cerebral cortex. *J. Neurosci.* **19,** 7881–7888.

Letinic, K., Zoncu, R., and Rakic, P. (2002). Origin of GABAergic neurons in the human neocortex [see comment]. *Nature* **417,** 645–649.

Levitt, P., Eagleson, K. L., and Powell, E. M. (2004). Regulation of neocortical interneuron development and the implications for neurodevelopmental disorders. *Trends Neurosci.* **27,** 400–406.

Lewine, J. D., Andrews, R., Chez, M., Patil, A. A., Devinsky, O., Smith, M., Kanner, A., Davis, J. T., Funke, M., Jones, G., Chong, B., Provencal, S., *et al.* (1999). Magnetoencephalographic patterns of epileptiform activity in children with regressive autism spectrum disorders [see comment]. *Pediatrics* **104,** 405–418.

Lewis, D. A., Hashimoto, T., and Volk, D. W. (2005). Cortical inhibitory neurons and schizophrenia. *Nat. Rev. Neurosci.* **6,** 312–324.

Liodis, P., Denaxa, M., Grigoriou, M., Akufo-Addo, C., Yanagawa, Y., and Pachnis, V. (2007). Lhx6 activity is required for the normal migration and specification of cortical interneuron subtypes. *J. Neurosci.* **27,** 3078–3089.

Lopez-Bendito, G., Lujan, R., Shigemoto, R., Ganter, P., Paulsen, O., and Molnar, Z. (2003). Blockade of GABA(B) receptors alters the tangential migration of cortical neurons. *Cereb. Corte.* **13,** 932–942.

Lopez-Bendito, G., Sanchez-Alcaniz, J. A., Pla, R., Borrell, V., Pico, E., Valdeolmillos, M., and Marin, O. (2008). Chemokine signaling controls intracortical migration and final distribution of GABAergic interneurons. *J. Neurosci.* **28,** 1613–1624.

LoTurco, J. J., Owens, D. F., Heath, M. J., Davis, M. B., and Kriegstein, A. R. (1995). GABA and glutamate depolarize cortical progenitor cells and inhibit DNA synthesis. *Neuron* **15,** 1287–1298.

Luhmann, H. J., and Prince, D. A. (1991). Postnatal maturation of the GABAergic system in rat neocortex. *J. Neurophys.* **65,** 247–263.

Ma, Y., Hu, H., Berrebi, A. S., Mathers, P. H., and Agmon, A. (2006). Distinct subtypes of somatostatin-containing neocortical interneurons revealed in transgenic mice. *J. Neurosci.* **26,** 5069–5082.

Manent, J. B., Demarque, M., Jorquera, I., Pellegrino, C., Ben-Ari, Y., Aniksztejn, L., and Represa, A. (2005). A noncanonical release of GABA and glutamate modulates neuronal migration. *J. Neurosci.* **25,** 4755–4765.

Manent, J. B., and Represa, A. (2007). Neurotransmitters and brain maturation: Early paracrine actions of GABA and glutamate modulate neuronal migration. *Neuroscientist* **13,** 268–279.

Marin, O., and Rubenstein, J. L. (2001). A long, remarkable journey: Tangential migration in the telencephalon. *Nat. Rev. Neurosci.* **2,** 780–790.

Marin, O., and Rubenstein, J. L. (2003). Cell migration in the forebrain. *Ann. Rev. Neurosci.* **26,** 441–483.

Markram, H., Toledo-Rodriguez, M., Wang, Y., Gupta, A., Silberberg, G., and Wu, C. (2004). Interneurons of the neocortical inhibitory system. *Nat. Rev. Neurosci.* **5,** 793–807.

Martina, M., Royer, S., and Pare, D. (2001). Cell-type-specific GABA responses and chloride homeostasis in the cortex and amygdala. *J. Neurophys.* **86,** 2887–2895.

McCabe, A. K., Chisholm, S. L., Picken-Bahrey, H. L., and Moody, W. J. (2006). The self-regulating nature of spontaneous synchronized activity in developing mouse cortical neurones. *J. Physiol.* **577,** 155–167.

McCabe, A. K., Easton, C. R., Lischalk, J. W., and Moody, W. J. (2007). Roles of glutamate and GABA receptors in setting the developmental timing of spontaneous synchronized activity in the developing mouse cortex. *Dev. Neurobio.* **67,** 1574–1588.

McConnell, S. K. (1988). Fates of visual cortical neurons in the ferret after isochronic and heterochronic transplantation. *J. Neurosci.* **8,** 945–974.

McConnell, S. K., and Kaznowski, C. E. (1991). Cell cycle dependence of laminar determination in developing neocortex. *Science* **254,** 282–285.

Merkle, F. T., Mirzadeh, Z., and Alvarez-Buylla, A. (2007). Mosaic organization of neural stem cells in the adult brain. *Science* **317,** 381–384.

Miller, M. T., Stromland, K., Ventura, L., Johansson, M., Bandim, J. M., and Gillberg, C. (2005). Autism associated with conditions characterized by developmental errors in early embryogenesis: A mini review. *Int. J. Dev. Neurosci.* **23,** 201–219.

Mission, J. P., Takahashi, T., and Caviness, V. S., Jr. (1991). Ontogeny of radial and other astroglial cells in murine cerebral cortex. *GLIA* **4,** 138–148.

Miyoshi, G., Butt, S. J., Takebayashi, H., and Fishell, G. (2007). Physiologically distinct temporal cohorts of cortical interneurons arise from telencephalic Olig2-expressing precursors. *J. Neurosci.* **27,** 7786–7798.

Molnar, G., Olah, S., Komlosi, G., Fule, M., Szabadics, J., Varga, C., Barzo, P., and Tamas, G. (2008). Complex events initiated by individual spikes in the human cerebral cortex. *Plos Biology* **6,** e222.

Molyneaux, B. J., Arlotta, P., Hirata, T., Hibi, M., and Macklis, J. D. (2005). Fezl is required for the birth and specification of corticospinal motor neurons. *Neuron* **47,** 817–831.

Monyer, H., and Markram, H. (2004). Interneuron Diversity series: Molecular and genetic tools to study GABAergic interneuron diversity and function. *Trends Neurosci.* **27,** 90–97.

Moretti, P., and Zoghbi, H. Y. (2006). MeCP2 dysfunction in Rett syndrome and related disorders. *Curr. Opin. Genet. Dev.* **16,** 276–281.

Mueller, A. L., Taube, J. S., and Schwartzkroin, P. A. (1984). Development of hyperpolarizing inhibitory postsynaptic potentials and hyperpolarizing response to gamma-aminobutyric acid in rabbit hippocampus studied *in vitro*. *J. Neurosci.* **4,** 860–867.

Nakagawa, Y., Johnson, J. E., and O'Leary, D. D. (1999). Graded and areal expression patterns of regulatory genes and cadherins in embryonic neocortex independent of thalamocortical input. *J. Neurosci.* **19,** 10877–10885.

Nery, S., Fishell, G., and Corbin, J. G. (2002). The caudal ganglionic eminence is a source of distinct cortical and subcortical cell populations. *Nat. Neurosci.* **5,** 1279–1287.

Nicol, X., Voyatzis, S., Muzerelle, A., Narboux-Neme, N., Sudhof, T. C., Miles, R., and Gaspar, P. (2007). cAMP oscillations and retinal activity are permissive for ephrin signaling during the establishment of the retinotopic map. *Nat. Neurosci.* **10,** 340–347.

Nobrega-Pereira, S., Kessaris, N., Du, T., Kimura, S., Anderson, S. A., and Marín, O. (2008). Postmitotic Nkx2–1 controls the migration of telencephalic interneurons by direct repression of guidance receptors [see comment]. *Neuron* **59,** 733–745.

O'Leary, D. D., Chou, S. J., and Sahara, S. (2007). Area patterning of the mammalian cortex. *Neuron* **56,** 252–269.

O'Leary, D. D., and Sahara, S. (2008). Genetic regulation of arealization of the neocortex. *Curr. Opin. Neurobiol.* **18,** 90–100.

O'Leary, D. D., and Stanfield, B. B. (1989). Selective elimination of axons extended by developing cortical neurons is dependent on regional locale: Experiments utilizing fetal cortical transplants. *J. Neurosci.* **9,** 2230–2246.

Obata, K., Oide, M., and Tanaka, H. (1978). Excitatory and inhibitory actions of GABA and glycine on embryonic chick spinal neurons in culture. *Brain Res.* **144,** 179–184.

Olsson, M., Bjorklund, A., and Campbell, K. (1998). Early specification of striatal projection neurons and interneuronal subtypes in the lateral and medial ganglionic eminence. *Neuroscience* **84,** 867–876.

Olsson, M., Campbell, K., Wictorin, K., and Bjorklund, A. (1995). Projection neurons in fetal striatal transplants are predominantly derived from the lateral ganglionic eminence. *Neuroscience* **69,** 1169–1182.

Owens, D. F., Boyce, L. H., Davis, M. B., and Kriegstein, A. R. (1996). Excitatory GABA responses in embryonic and neonatal cortical slices demonstrated by gramicidin perforated-patch recordings and calcium imaging. *J. Neurosci.* **16,** 6414–6423.

Owens, D. F., and Kriegstein, A. R. (1998). Patterns of intracellular calcium fluctuation in precursor cells of the neocortical ventricular zone. *J. Neurosci.* **18,** 5374–5388.

Parnavelas, J. G. (2000). The origin and migration of cortical neurones: New vistas. *Trends Neurosci.* **23,** 126–131.

Peinado, A. (2000). Traveling slow waves of neural activity: A novel form of network activity in developing neocortex. *J. Neurosci.* **20,** RC54.

Petilla Interneuron Nomenclature Group, Ascoli, G. A., Alonso-Nanclares, L., Anderson, S. A., Barrionuevo, G., Benavides-Piccione, R., Burkhalter, A., Buzsaki, G., Cauli, B., Defelipe, J., Fairén, A., Feldmeyer, D., Fishell, G., *et al.* (2008). Petilla terminology: Nomenclature of features of GABAergic interneurons of the cerebral cortex. *Nat. Rev. Neurosci.* **9,** 557–568.

Petryniak, M. A., Potter, G. B., Rowitch, D. H., and Rubenstein, J. L. (2007). Dlx1 and Dlx2 control neuronal versus oligodendroglial cell fate acquisition in the developing forebrain. *Neuron* **55,** 417–433.

Pieper, A. A., Wu, X., Han, T. W., Estill, S. J., Dang, Q., Wu, L. C., Reece-Fincanon, S., Dudley, C. A., Richardson, J. A., Brat, D. J., and McKnight, S. L. (2005). The neuronal PAS domain protein 3 transcription factor controls FGF-mediated adult hippocampal neurogenesis in mice. *Proc. Natl. Acad. Sci. USA* **102,** 14052–14057.

Pleasure, S. J., Anderson, S., Hevner, R., Bagri, A., Marin, O., Lowenstein, D. H., and Rubenstein, J. L. (2000). Cell migration from the ganglionic eminences is required for the development of hippocampal GABAergic interneurons. *Neuron* **28,** 727–740.

Polleux, F., Whitford, K. L., Dijkhuizen, P. A., Vitalis, T., and Ghosh, A. (2002). Control of cortical interneuron migration by neurotrophins and PI3-kinase signaling. *Development* **129,** 3147–3160.

Poluch, S., and Juliano, S. L. (2007). A normal radial glial scaffold is necessary for migration of interneurons during neocortical development. *GLIA* **55,** 822–830.

Pombal, M. A., and Puelles, L. (1999). Prosomeric map of the lamprey forebrain based on calretinin immunocytochemistry, Nissl stain, and ancillary markers. *J. Comp. Neurol.* **414,** 391–422.

Porter, J. T., Cauli, B., Staiger, J. F., Lambolez, B., Rossier, J., and Audinat, E. (1998). Properties of bipolar VIPergic interneurons and their excitation by pyramidal neurons in the rat neocortex. *Eur. J. Neurosci.* **10,** 3617–3628.

Porteus, M. H., Bulfone, A., Liu, J. K., Puelles, L., Lo, L. C., and Rubenstein, J. L. (1994). DLX-2, MASH-1, and MAP-2 expression and bromodeoxyuridine incorporation define molecularly distinct cell populations in the embryonic mouse forebrain. *J. Neurosci.* **14,** 6370–6383.

Pouille, F., and Scanziani, M. (2001). Enforcement of temporal fidelity in pyramidal cells by somatic feed-forward inhibition. *Science* **293,** 1159–1163.

Puelles, L. (2001). Brain segmentation and forebrain development in amniotes. *Brain Res. Bulletin* **55,** 695–710.

Puelles, L., Kuwana, E., Puelles, E., Bulfone, A., Shimamura, K., Keleher, J., Smiga, S., and Rubenstein, J. L. (2000). Pallial and subpallial derivatives in the embryonic chick and mouse telencephalon, traced by the expression of the genes Dlx-2, Emx-1, Nkx-2.1, Pax-6, and Tbr-1. *J. Comp. Neurol.* **424,** 409–438.

Puelles, L., and Rubenstein, J. L. (2003). Forebrain gene expression domains and the evolving prosomeric model. *Trends Neurosci.* **26,** 469–476.

Rakic, P. (1974). Neurons in rhesus monkey visual cortex: Systematic relation between time of origin and eventual disposition. *Science* **183,** 425–427.

Rakic, P. (1988). Specification of cerebral cortical areas. *Science* **241,** 170–176.

Rakic, P. (1990). Principles of neural cell migration. *Experientia* **46,** 882–891.

Rallu, M., Corbin, J. G., and Fishell, G. (2002a). Parsing the prosencephalon. *Nat. Rev. Neurosci.* **3,** 943–951.

Rallu, M., Machold, R., Gaiano, N., Corbin, J. G., McMahon, A. P., and Fishell, G. (2002b). Dorsoventral patterning is established in the telencephalon of mutants lacking both Gli3 and Hedgehog signaling. *Development* **129,** 4963–4974.

Rubenstein, J. L., Martinez, S., Shimamura, K., and Puelles, L. (1994). The embryonic vertebrate forebrain: The prosomeric model. *Science* **266,** 578–580.

Rubenstein, J. L., Shimamura, K., Martinez, S., and Puelles, L. (1998). Regionalization of the prosencephalic neural plate. *Annu. Rev. Neurosci.* **21,** 445–477.

Rudy, B., and McBain, C. J. (2001). Kv3 channels: Voltage-gated K+ channels designed for high-frequency repetitive firing. *Trends Neurosci.* **24,** 517–526.

Rymar, V. V., and Sadikot, A. F. (2007). Laminar fate of cortical GABAergic interneurons is dependent on both birthdate and phenotype. *J. Comp. Neurol.* **501,** 369–380.

Shen, Q., Wang, Y., Dimos, J. T., Fasano, C. A., Phoenix, T. N., Lemischka, I. R., Ivanova, N. B., Stifani, S., Morrisey, E. E., and Temple, S. (2006). The timing of cortical neurogenesis is encoded within lineages of individual progenitor cells [see comment]. *Nat. Neurosci.* **9,** 743–751.

Shimamura, K., Hartigan, D. J., Martinez, S., Puelles, L., and Rubenstein, J. L. (1995). Longitudinal organization of the anterior neural plate and neural tube. *Development* **121,** 3923–3933.

Shimamura, K., Martinez, S., Puelles, L., and Rubenstein, J. L. (1997). Patterns of gene expression in the neural plate and neural tube subdivide the embryonic forebrain into transverse and longitudinal domains. *Dev. Neurosci.* **19,** 88–96.

Smart, I. H. (1976). A pilot study of cell production by the ganglionic eminences of the developing mouse brain. *J. Anat.* **121,** 71–84.

Somogyi, P., and Klausberger, T. (2005). Defined types of cortical interneurone structure space and spike timing in the hippocampus. *J. Physiol.* **562,** 9–26.

Somogyi, P., Tamas, G., Lujan, R., and Buhl, E. H. (1998). Salient features of synaptic organisation in the cerebral cortex. *Brain Res. - Brain Res. Rev.* **26,** 113–135.

Sousa, V. H., Miyoshi, G., Hjerling-Leffler, J., Karayannis, T. and Fishell, G. (2009). Characterization of Nkx6-2-Derived Neocortical Interneuron Lineages. *Cereb Cortex.* In Press.

Stenman, J., Toresson, H., and Campbell, K. (2003a). Identification of two distinct progenitor populations in the lateral ganglionic eminence: Implications for striatal and olfactory bulb neurogenesis. *J. Neurosci.* **23,** 167–174.

Stenman, J. M., Wang, B., and Campbell, K. (2003b). Tlx controls proliferation and patterning of lateral telencephalic progenitor domains. *J. Neurosci.* **23,** 10568–10576.

Stuhmer, T., Anderson, S. A., Ekker, M., and Rubenstein, J. L. (2002). Ectopic expression of the Dlx genes induces glutamic acid decarboxylase and Dlx expression. *Development* **129,** 245–252.

Stumm, R. K., Zhou, C., Ara, T., Lazarini, F., Dubois-Dalcq, M., Nagasawa, T., Hollt, V., and Schulz, S. (2003). CXCR4 regulates interneuron migration in the developing neocortex. *J. Neurosci.* **23,** 5123–5130.

Sturrock, R. R., and Smart, I. H. (1980). A morphological study of the mouse subependymal layer from embryonic life to old age. *J. Anat.* **130,** 391–415.

Sugino, K., Hempel, C. M., Miller, M. N., Hattox, A. M., Shapiro, P., Wu, C., Huang, Z. J., and Nelson, S. B. (2006). Molecular taxonomy of major neuronal classes in the adult mouse forebrain [see comment][erratum appears in *Nat. Neurosci.* 2006 Feb;9 (2):292]. *Nat. Neurosci.* **9,** 99–107.

Sur, M., and Rubenstein, J. L. (2005). Patterning and plasticity of the cerebral cortex. *Science* **310,** 805–810.

Sussel, L., Marin, O., Kimura, S., and Rubenstein, J. L. (1999). Loss of Nkx2.1 homeobox gene function results in a ventral to dorsal molecular respecification within the basal telencephalon: Evidence for a transformation of the pallidum into the striatum. *Development* **126,** 3359–3370.

Taglialatela, P., Soria, J. M., Caironi, V., Moiana, A., and Bertuzzi, S. (2004). Compromised generation of GABAergic interneurons in the brains of Vax1-/- mice. *Development* **131,** 4239–4249.

Tamamaki, N., Fujimori, K. E., and Takauji, R. (1997). Origin and route of tangentially migrating neurons in the developing neocortical intermediate zone. *J. Neurosci.* **17,** 8313–8323.

Tiveron, M. C., Rossel, M., Moepps, B., Zhang, Y. L., Seidenfaden, R., Favor, J., Konig, N., and Cremer, H. (2006). Molecular interaction between projection neuron precursors and invading interneurons via stromal-derived factor 1 (CXCL12)/CXCR4 signaling in the cortical subventricular zone/intermediate zone. *J. Neurosci.* **26,** 13273–13278.

Toresson, H., and Campbell, K. (2001). A role for Gsh1 in the developing striatum and olfactory bulb of Gsh2 mutant mice. *Development* **128,** 4769–4780.

Toresson, H., Potter, S. S., and Campbell, K. (2000). Genetic control of dorsal-ventral identity in the telencephalon: Opposing roles for Pax6 and Gsh2. *Development* **127,** 4361–4371.

Tyzio, R., Holmes, G. L., Ben-Ari, Y., and Khazipov, R. (2007). Timing of the developmental switch in GABA(A) mediated signaling from excitation to inhibition in CA3 rat hippocampus using gramicidin perforated patch and extracellular recordings. *Epilepsia* **48,** 96–105.

Valcanis, H., and Tan, S. S. (2003). Layer specification of transplanted interneurons in developing mouse neocortex. *J. Neurosci.* **23,** 5113–5122.

Vincent, J. B., Horike, S. I., Choufani, S., Paterson, A. D., Roberts, W., Szatmari, P., Weksberg, R., Fernandez, B., and Scherer, S. W. (2006). An inversion inv(4)(p12-p15.3) in autistic siblings implicates the 4p GABA receptor gene cluster. *J. Med. Genet.* **43,** 429–434.

Waclaw, R. R., Allen, Z. J., 2nd, Bell, S. M., Erdelyi, F., Szabo, G., Potter, S. S., and Campbell, K. (2006). The zinc finger transcription factor Sp8 regulates the generation and diversity of olfactory bulb interneurons. *Neuron* **49,** 503–516.

Wheless, J. W., and Kim, H. L. (2002). Adolescent seizures and epilepsy syndromes. *Epilepsia* **43,** 33–52.

Wichterle, H., Garcia-Verdugo, J. M., Herrera, D. G., and Alvarez-Buylla, A. (1999). Young neurons from medial ganglionic eminence disperse in adult and embryonic brain. *Nat. Neurosci.* **2,** 461–466.

Wichterle, H., Turnbull, D. H., Nery, S., Fishell, G., and Alvarez-Buylla, A. (2001). In utero fate mapping reveals distinct migratory pathways and fates of neurons born in the mammalian basal forebrain. *Development* **128,** 3759–3771.

Wilson, S. W., and Houart, C. (2004). Early steps in the development of the forebrain. *Dev. Cell* **6,** 167–181.

Wilson, S. W., and Rubenstein, J. L. (2000). Induction and dorsoventral patterning of the telencephalon. *Neuron* **28,** 641–651.

Wonders, C. P., and Anderson, S. A. (2006). The origin and specification of cortical interneurons. *Nat. Rev. Neurosci.* **7,** 687–696.

Wonders, C. P., Taylor, L., Welagen, J., Mbata, I. C., Xiang, J. Z., and Anderson, S. A. (2008). A spatial bias for the origins of interneuron subgroups within the medial ganglionic eminence. *Dev. Bio.* **314,** 127–136.

Woo, N. H., and Lu, B. (2006). Regulation of cortical interneurons by neurotrophins: From development to cognitive disorders. *Neurosci.* **12,** 43–56.

Xu, Q., Cobos, I., De La Cruz, E., Rubenstein, J. L., and Anderson, S. A. (2004). Origins of cortical interneuron subtypes. *J. Neurosci.* **24,** 2612–2622.

Xu, Q., Tam, M., and Anderson, S. A. (2008). Fate mapping Nkx2.1-lineage cells in the mouse telencephalon. *J. Comp. Neurol.* **506,** 16–29.

Yun, K., Garel, S., Fischman, S., and Rubenstein, J. L. (2003). Patterning of the lateral ganglionic eminence by the Gsh1 and Gsh2 homeobox genes regulates striatal and olfactory bulb histogenesis and the growth of axons through the basal ganglia. *J. Comp. Neurol.* **461,** 151–165.

Yun, K., Potter, S., and Rubenstein, J. L. (2001). Gsh2 and Pax6 play complementary roles in dorsoventral patterning of the mammalian telencephalon. *Development* **128,** 193–205.

Yuste, R., Peinado, A., and Katz, L. C. (1992). Neuronal domains in developing neocortex. *Science* **257,** 665–669.

CHAPTER FOUR

Transcriptional Networks in the Early Development of Sensory–Motor Circuits

Jeremy S. Dasen

Contents

Smilow Neuroscience Program, Department of Physiology and Neuroscience, New York University School of Medicine, New York, New York, USA

Current Topics in Developmental Biology, Volume 87
ISSN 0070-2153, DOI: 10.1016/S0070-2153(09)01204-6

Abstract

The emergence of coordinated locomotor behaviors in vertebrates relies on the establishment of selective connections between discrete populations of neurons present in the spinal cord and peripheral nervous system. The assembly of the circuits necessary for movement presumably requires the generation of many unique cell types to accommodate the intricate connections between motor neurons, sensory neurons, interneurons, and muscle. The specification of diverse neuronal subtypes is mediated largely through networks of transcription factors that operate within progenitor and postmitotic cells. Selective patterns of transcription factor expression appear to define the cell-type-specific cellular programs that govern the axonal guidance decisions and synaptic specificities of neurons, and may lay the foundation through which innate motor behaviors are genetically predetermined. Recent studies on the developmental programs that specify two highly diverse neuronal classes—spinal motor neurons and proprioceptive sensory neurons—have provided important insights into the molecular strategies used in the earliest phases of locomotor circuit assembly. This chapter reviews progress toward elucidating the early transcriptional networks that define neuronal identity in the locomotor system, focusing on the pathways controlling the specific connections of motor neurons and sensory neurons in the formation of simple reflex circuits.

1. Introduction

Many organisms are born with a set of innate behaviors that have evolved so that they can confront the challenges imposed by their specific environments. The neural circuits controlling basic motor behaviors such as feeding, breathing, and walking are often functional at the time of birth, independent of any prior interaction with the external world. These genetically hard-wired circuits can be essential for survival by imprinting behaviors such as the predator escape response—a system where sensory input must be tightly linked to motor output. The identification of the substrates for simple and complex innate behaviors has been a major challenge.

There is emerging evidence that stereotyped patterns of movement can be programmed through the actions of a few key regulatory genes, neurons, and microcircuits. In *Drosophila* gender-specific courtship behaviors are specified by a transcription factor encoded by the *fruitless* gene, which is sufficient to interconvert a specific pattern of mating behavior between males and females (Demir and Dickson, 2005). In *Caenorhabditis elegans*, the neural substrates controlling innate patterns of motor behavior are comprised of a relatively small number of anatomically well-defined groups of neurons (Hobert, 2003). Whether similar master gene regulators, or discrete neural assemblies, function to program the behavioral outputs of the vertebrate motor system remain to be determined. Presumably such a factor, or group of factors, would need to function in distinct classes of interconnected neuronal subtypes that by other criteria might be considered dissimilar.

The problem of defining behaviorally relevant circuits in vertebrate nervous systems is confounded by the shear volume of neurons, and the relative complexity and number of synaptic connections. The spinal cord and hindbrain have provided tractable model systems for defining the neural circuits necessary for basic motor functions such as breathing and walking, and contain the sensory feedback systems that are required for reflex responses and locomotor adaptation (reviewed in Goulding and Pfaff, 2005; Kiehn and Butt, 2003). As a system for studying locomotor behaviors, the spinal cord has an advantage in that the anatomy of the system is relatively well defined and the sensory inputs and motor outputs are accessible and quantifiable. One successful approach to the study of the assembly of locomotor circuits has been to define the embryonic programs that contribute to the identity and connectivity of the cells within the circuit, to try to link control of synaptic specificity with the emergence of a defined behavior.

In this chapter, I review our current understanding of the genetic programs which control the specification of motor neurons and sensory neurons in the vertebrate spinal cord and peripheral nervous system. Emphasis will be placed upon the transcriptional networks which dictate the early identity of these two neuronal classes, and on recent advances that have enriched our understanding of the general principles of circuit assembly. Activity-dependent steps in the wiring of locomotor circuits will not be addressed as this aspect has been the subject of recent reviews (Hanson *et al.*, 2008; Ladle *et al.*, 2007). The potential mechanisms that may contribute to the assembly of sensory–motor circuits will be explored, with a focus on the formation of monosynaptic stretch-reflex circuits, a collection of neural circuits that are critical for coordinated movement.

2. Intrinsic Programs Controlling Neuronal Fate Specification in the Ventral Spinal Cord

In broad terms, the final output of spinal circuit activity is conceptually simple: the activation of specific muscles in the periphery. But in order for basic motor commands to be smoothly executed, spinal circuits must be sufficiently fined tuned to activate only a small subset of the hundreds of unique muscle groups in a specific order. The first and most critical aspect in the formation of these circuits is that motor axons be able to navigate toward and select their peripheral muscle targets with fidelity and precision. All of the subsequent steps in motor neuron connectivity, such as descending inputs from higher brain centers, local interneuron, and sensory neuron connectivity are constrained by the peripheral connections made by motor neurons. Therefore connection between a motor neuron and the muscle it innervates is a core element in defining synaptic specificity during locomotor circuit assembly.

The problem of establishing appropriate sets of connections between motor neurons and muscles is at its most challenging in the vertebrate limb, which contains dozens of anatomically and functionally distinct muscle groups in most tetrapods (Greene, 1963; Sullivan, 1962). The highly stereotypic nature of developing motor axonal projections within the limb, and the organization of motor neuron cell bodies into topographic maps, led to the idea that motor neurons possess intrinsic identities that define target specificity (Jessell, 2000; Landmesser, 2001). The patterns of motor neuron connectivity in the limb require that a large number of unique identities are generated during development, to accommodate the targeting of each of the approximately 50 muscle groups in the limb. As described below, the diversification of motor neuron subtypes, likely many CNS neurons, appears to be mediated through the integration of two distinct signaling systems that operate along the dorsoventral and rostrocaudal axis of the neural tube.

2.1. Control of neuronal class identity in the ventral spinal cord

Many of the neural circuits which provide the central drive for activation of motor neurons reside within ventral spinal cord. The ventral spinal cord consists of several classes of neurons including motor neurons and multiple types of local circuit interneurons (Goulding and Pfaff, 2005). The identities of these classes are established largely through the actions of the secreted signaling molecule Sonic hedgehog (Shh), which is secreted from the notochord and floor plate and acts in a graded manner to pattern progenitor identities along the dorsoventral axis of the neural tube (Dessaud *et al.*, 2008). A primary function of Shh signaling is to control the expression patterns of homeodomain and helix-loop-helix classes of transcription factors, most of which act as transcriptional repressors (Briscoe *et al.*, 2000; Muhr *et al.*, 2001; Novitch *et al.*, 2001). Two classes of Shh-regulated genes have been defined; class I genes are expressed more dorsally and are repressed by Shh, while class II genes are found ventrally and are activated in response to graded Shh signaling (Briscoe *et al.*, 2000). These initial patterns of expression are subsequently refined through mutually cross-repressive interactions that occur between pairs of transcription factors. As a consequence of these repressor interactions, each domain expresses a unique transcription factor code that specifies progenitor identities, including motor neuron (pMN) and interneuron (p0, p1, p2, p3) precursors (Fig. 4.1A).

Each cardinal progenitor domain gives rise to specific class of postmitotic neuron characterized by a set of common features such as axonal projection pattern, settling position, synaptic partners, and physiological properties. Like the transcriptional codes present in neural progenitors, postmitotic neuronal classes can also be defined by their unique and combinatorial expression of specific transcription factors. For example, the progenitor domain that gives rise to motor neurons expresses the basic helix-loop-helix (bHLH) protein Olig2 and Nkx6 homeodomain proteins (Briscoe *et al.*, 2000; Novitch *et al.*, 2001). After leaving the cell cycle these newly postmitotic motor neurons are characterized by the expression of a distinct set of homeodomain including factors Hb9, Isl1/2, and Lhx3/4 (Fig. 4.1A) (Tanabe *et al.*, 1998; Tsuchida *et al.*, 1994).

Genetic loss-of-function studies in mouse have established critical roles for these class-specific transcription factors in neuronal subtype specification and have also revealed their essential roles in locomotor circuit assembly and function. In mice lacking the V0 interneuron transcription factor Dbx1, mice fail to form appropriate inhibitory connections between the two sides of the spinal cord and as a consequence the normal alternation of left and right motor outputs of the spinal cord are disrupted (Lanuza *et al.*, 2004). Thus, genetic analyses of class-specific transcription factors have revealed the general mechanisms governing the specification of diverse neuronal classes in the spinal cord, and have also provided important insights into how individual classes of neurons contribute to locomotor circuit function.

2.2. Defining positional identities along the rostrocaudal axis of the spinal cord

Although transcriptional networks mediated through the dorsoventral signaling systems define the identity of several types of neurons, other programs are presumably necessary for the further diversification of neurons within a given class. Classical anatomical and physiological studies indicate that much of the variation of neuronal subtypes in the spinal cord occurs as a function of their position along the rostrocaudal axis. For example, the neuronal circuits responsible for the rhythmic firing of motor neurons during stereotyped behaviors such as walking are located at defined positions along the rostrocaudal axis (Kiehn and Butt, 2003). These specialized limb-controlling circuits (known as central pattern generators—CPGs) likely reflect the programming of neuronal identities at specific rostrocaudal positions.

The mechanisms controlling neural diversification along the rostrocaudal axis of the spinal cord are best understood for motor neurons. While all

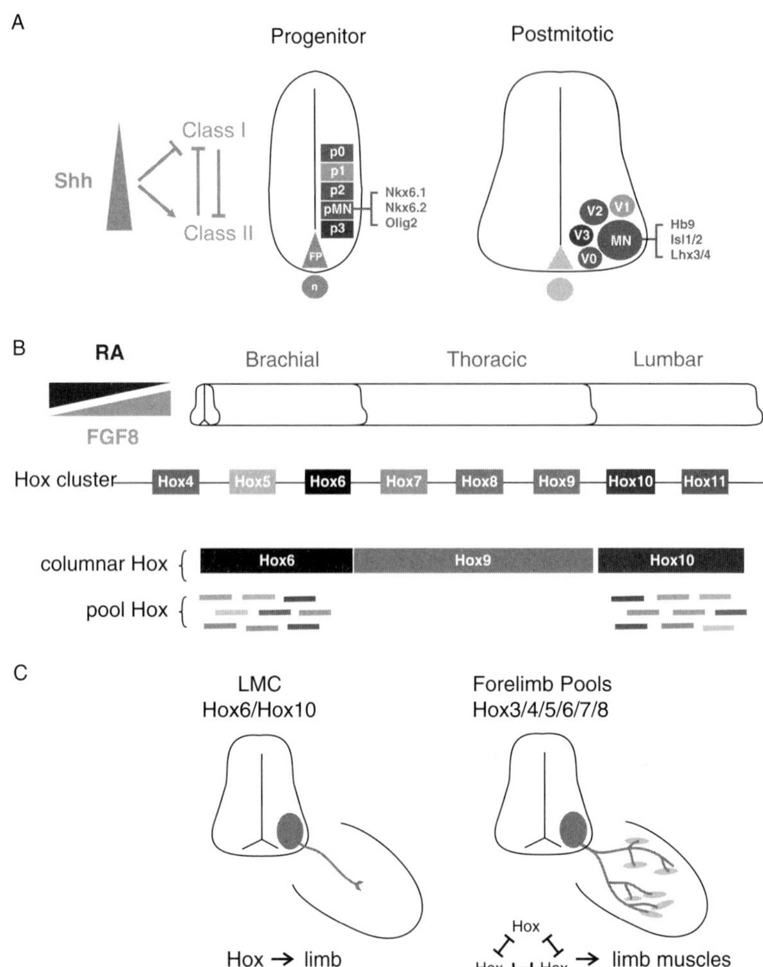

Figure 4.1 Steps in the generation of neuronal subtypes in the ventral spinal cord. (A) Along the dorsoventral axis of the neural tube, several classes of neurons are generated in response to the graded activities of the secreted protein Sonic hedgehog (Shh) which induces the patterned expression of transcription factors in ventral progenitor cells. "Class I" transcription factors are induced by Shh while "Class II" factors are repressed. Selective cross-repressive interactions between these two classes of transcription factors sharpen the boundaries between progenitor domains (see Briscoe *et al.*, 2000). In addition, retinoic acid (RA) from the paraxial mesoderm and fibroblast growth factor (FGF) signaling also influence the pattern of transcription factor expression in neural tube progenitors (not shown). Each of these progenitor domains gives rise to postmitotic neurons, including motor neurons and several classes of inhibitory and excitatory local circuit interneurons. (B) Along the rostrocaudal axis of the neural tube, opposing FGF and RA gradients induce the expression of chromosomally linked *Hox* genes. *Hox* genes located at one end of the cluster are induced more rostrally in response to high levels of RA while genes at the opposite end are expressed caudally

motor neurons are characterized by certain common features, such as their cholinergic mode of neurotransmission, the extension of axons outside the CNS, and expression of class-specific transcription factors, they also acquire intrinsic features which differentiate them from one another. At limb levels, three main levels of motor neuron organization have been defined based on studies of their position within the spinal cord and pattern of connectivity, with each successive step governing a distinct aspect of motor axonal projection pattern (Fig. 4.2). Shortly after leaving the cell cycle, a subset of motor neurons acquire a lateral motor neuron columnar (LMC) identity that directs motor axons toward the limb. Once motor axons reach the base of the limb, the specification of divisional identities within the LMC directs motor axons ventrally or dorsally upon entering the limb mesenchyme. Finally, the specification of a motor pool identity appears to confer motor neurons with the ability to project toward and form precise connections with individual limb muscle targets (Landmesser, 1978a, 2001).

Each of these sequential phases of limb innervation pattern can be linked to a genetic program defined by transcriptional networks which utilize members of the evolutionarily conserved *Hox* gene family. The generation of segmentally restricted motor neuron columnar subtypes are specified by rostrocaudal signaling gradients of fibroblast growth factors (FGFs) and retinoic acid (RA) that establish regional domains of Hox transcription factor expression and activity (Fig. 4.1B) (Bel-Vialar *et al.*, 2002; Liu *et al.*, 2001). The FGF and RA gradients set up an initial pattern of *Hox* expression in motor neurons that is subsequently refined through cross-repressive interactions that occur between specific pairs of Hox proteins (Dasen *et al.*, 2003). As a consequence of these early inductive signals and cross-regulatory interactions, Hox expression patterns segregate with and determine the identity of distinct motor neuron subtypes (Dasen *et al.*, 2003, 2005). For example, at limb levels Hox6 gene paralogs (*Hoxa6* and *Hoxc6*) specify forelimb LMC neuron fate, while Hox10 genes (*Hoxa10*, *Hoxc10*, and *Hoxd10*) specify hindlimb LMC neurons (Dasen *et al.*, 2003; Shah *et al.*, 2004; Wu *et al.*, 2008).

A more intricate transcriptional network, built from nearly two dozen Hox proteins, imposes discrete motor pool identities in LMC neurons

in response to elevated levels of FGF. In motor neurons, Hox patterns are further refined through cross-repressive interactions, giving rise to specific patterns of Hox expression in motor columns and pools. (C) Role of Hox proteins genes in motor neuron innervation patterns in the limb. One set of Hox factors expressed at limb levels of the spinal cord is involved in establishing the lateral motor neuron columnar (LMC) identity which directs motor axons toward the limb. Within this columnar group, an additional network of 8–10 additional Hox factors appear to be involved in the specification of motor pool identities and control programs which direct motor axons to specific muscle targets.

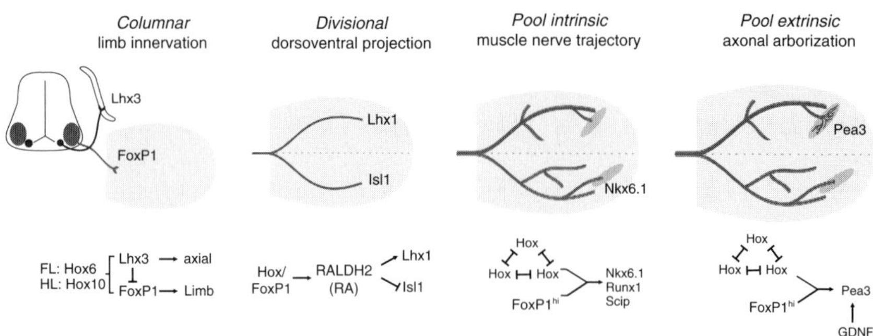

Figure 4.2 Sequential steps in the transcriptional control of limb innervation patterns. Two motor columns are present at limb levels of the spinal cord, one projecting dorsally toward axial muscle the other projecting ventrally into the limb. The LIM homeodomain factor Lhx3 specifies the identity of dorsally projecting neurons, while FoxP1 appears to establish the early limb innervation program. Hox6 proteins specify forelimb (FL) LMC neurons and Hox10 hindlimb (HL) neurons. Upon entering the limb, motor axons follow either a dorsal or ventral trajectory. The dorsoventral decision operates as a consequence of the pattern of LIM homeodomain expression setup by a local source of retinoids, provided by *RALDH2* activity. The guidance of axons to specific muscle is determined by a Hox network operating in the presence of high levels of FoxP1 that establishes the transcriptional identity of motor pools. Late aspects of this program appear to require signaling from the periphery, and these peripheral signals (e.g., GDNF) control expression of ETS transcription factors, such as Pea3. One aspect of Pea3 function in motor neurons is to control muscle-specific patterns of axonal arborization.

(Dasen *et al.*, 2005). Within motor pools Hox proteins appear to act as the primary determinants of motor neuron connectivity since changing the profile of Hox expression in specific pools, through misexpression or *Hox* gene mutation, results in changes in the pattern of muscle innervation (Dasen *et al.*, 2005; Tarchini *et al.*, 2005; Tiret *et al.*, 1998; Wu *et al.*, 2008). As described in the following sections, this core Hox network appears to direct motor innervation patterns by activating a diverse array of downstream transcription factors and cell surface receptors.

2.3. Transcriptional networks in motor neuron columnar and pool identities

Despite evidence that *Hox* genes are critical for motor neuron differentiation, their activities alone cannot account for all aspects of motor neuron diversity. *Hox* genes are global regulators of tissue patterning throughout development (Kmita and Duboule, 2003), and within the locomotor system are expressed by multiple classes of neurons (Belting *et al.*, 1998; Ensini *et al.*, 1998). The ability of *Hox* genes to selectively control columnar, divisional, and pool identities of motor neurons appears to require the actions of additional transcription factors which act both in parallel and downstream of Hox function.

The network of Hox proteins that drives motor neuron diversification depends on the actions of a single accessory factor, the forkhead class homeodomain protein FoxP1. Genetic inactivation of the *Foxp1* gene in mice erases all of the Hox-dependent steps of LMC motor neurons differentiation and motor neuron identities revert to an ancestral state, consisting of two continuous motor columns (Dasen *et al.*, 2008; Rousso *et al.*, 2008). In *Foxp1* mutants, all known molecular feature of LMC neuron diversification are deteriorated, including the expression of columnar, divisional, and pool-restricted transcription factors and guidance receptors. Remarkably, the overall pattern of axonal projections into limb is well preserved in these mutants although individual motor neurons appear to select projection pathways in the limb in a stochastic manner (Fig. 4.3A). More generally, these observations suggest that the axonal projection pattern observed in the limb may be in part set up by the limb itself, independent of expression of guidance receptors on specific motor axons.

While these studies indicate that FoxP1 is required for Hox actions in motor neurons, it remains unclear whether FoxP1 is a bona fide cofactor for Hox proteins in motor neurons. In addition, how the Hox/FoxP1 network interacts with more globally acting motor neuron determinants such as LIM homeodomain proteins and Hb9, which are required in most motor neurons (Arber *et al.*, 1999; Thaler *et al.*, 1999, 2004), remains to be determined. Since Hox proteins are expressed by interneurons and sensory neurons, it is possible that additional cofactors control Hox-regulated aspects of neuronal identity and connectivity. One possibility is that the Hox network described for motor neurons functions in other classes of spinal neurons to help determine their synaptic specificities.

3. Guidance and Synaptic Specificity of Motor Axons Projecting into the Limb

While there is significant evidence that a Hox/FoxP1-based transcriptional network contributes to the diversity and connectivity of motor neuron subtypes, the specific molecular pathways by which this program contributes to the guidance of motor axons to their muscle targets is not well defined. Nevertheless, the actions of certain Hox proteins can be linked to the ability of motor neurons to innervate specific muscle targets through the control of a diverse repertoire of intermediate factors (Fig. 4.2). One early output of the Hox-based program controlling limb innervation is the establishment of a code of LIM homeodomain protein expression that

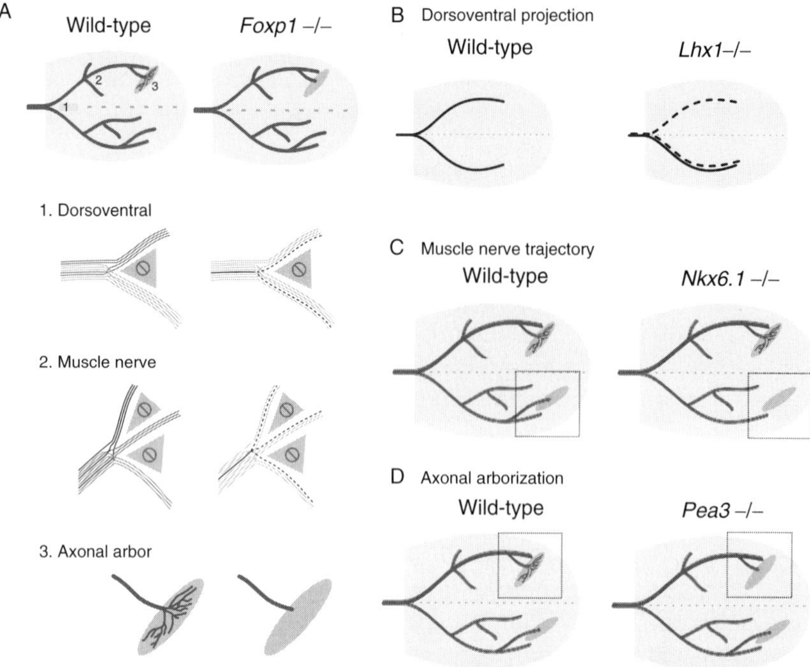

Figure 4.3 Genetic analysis of transcription factors involved in motor axon projections into the limb. (A) In *Foxp1* mutants, the Hox-dependent programs for limb-innervating motor neurons deteriorate, including the loss of pool-specific transcription factors and guidance molecules. The pattern of motor nerve branches is similar in wild-type and *Foxp1* mutant embryos, but the behavior of individual motor axons differs markedly at decision regions encountered in the limb mesenchyme. In *Foxp1* mutants, the axons of motor neurons are still forced to follow permissive pathways in the limb mesenchyme, but deprived of their divisional or pool identity, they select one of the permitted routes in an apparently stochastic manner. "No entry" signs depict zones of the limb mesenchyme from which motor axons are excluded, presumably by virtue of the expression of molecules that are nonpermissive for axon extension. Late aspects of pool differentiation such as muscle-specific axonal arborization patterns are also eroded. (B) In mice mutant for *Lhx1*, which is normally expressed by dorsally projecting lateral LMC motor neurons, the axons that normally project into the dorsal compartment of the limb appear to choose the selection of specific targets in random manner. This randomization is likely due to loss of EphA4 expression which is required to repel this population from ephrins expressed in the ventral limb mesenchyme. (C) At hindlimb levels, the motor nerve branches supplying the tibialis anterior and gracilis posterior muscles in the hindlimb derive from Nkx6.1+ motor pools. In *Nkx6.1* mutants, the axons that normally project to these muscles are rerouted to different muscle targets. (D) In *Pea3* mutants, motor axons from the motor pool expressing Pea3 are able to navigate to their appropriate muscle target. Once they reach their target, however, they are unable to arborize the muscle.

defines the dorsoventral projection choice of motor axon at the base of the limb. While an additional Hox program imparts a second layer of control that specifies the motor neuron pools targeting individual limb muscles.

3.1. LIM homeodomain proteins and axonal trajectories

The cell bodies of motor neurons that inhabit the LMC form a topographic map in the spinal cord such that cell body position is predictive of axonal projection pattern (Sharma and Belmonte, 2001). How the somatotopic organization of motor neurons emerges is best understood for the two divisions of the LMC. These divisional identities are established in a binary manner: motor neurons that settle in the medial half of the LMC innervate ventrally derived limb muscles, whereas motor neurons that settle in the lateral half of the LMC innervate dorsally derived limb muscles (Landmesser, 1978a; Tosney and Landmesser, 1985a,b). At a molecular level, the divisional organization of LMC neurons is revealed in the expression of a set of LIM homeodomain proteins (Tsuchida *et al.*, 1994). Laterally positioned LMC neurons express Lhx1 whereas medially positioned neurons express Isl1. This pattern of LIM homeodomain expression appears to be established through signaling between motor neurons just after they are born and can be linked to the activities of Hox proteins expressed in these subtypes.

At limb levels of the spinal cord, LMC neurons are characterized by expression of the *retinaldehyde dehydrogenase 2 (RALDH2)* gene which encodes an enzyme involved in retinoic acid synthesis. The selective expression of *RALDH2* in LMC neurons is controlled by the activities of Hox proteins expressed at limb levels of the spinal cord: Hox6 paralogs activate *RALDH2* expression at forelimb levels and Hox10 paralogs control its expression at hindlimb levels (Dasen *et al.*, 2003; Shah *et al.*, 2004). This action of Hox factors depends on the presence of accessory factor FoxP1, since in *Foxp1* mutants *RALDH2* is only weakly expressed (Dasen *et al.*, 2008; Rousso *et al.*, 2008). Whether the action of FoxP1 and Hox proteins in the control of *RALDH2* involves cooperative protein–protein interactions, or whether FoxP1 is merely an intermediary factor remains to be determined.

RALDH2 activity in LMC neurons generates a local source of retinoid signaling which appears to be responsible for establishing the pattern of LIM homeodomain expression in lateral and medial LMC neurons (Sockanathan and Jessell, 1998). At this stage, motor neurons can be characterized by a pattern of "inside–out" migration where later born lateral LMC neurons migrate past the early born medial LMC neurons. As later born LMC neurons migrate past the early born counterparts, RA provided by medial LMC neurons induces the expression of Lhx1 and specifies a lateral LMC fate. The induction of Lhx1 is also involved in restricting Isl1 expression to medial LMC neurons, thus helping to set up a binary code of LIM homeodomain expression within these two populations (Kania and Jessell, 2003; Sockanathan and Jessell, 1998).

How do the actions of LIM homeodomain proteins dictate the initial trajectory of motor axons as they enter the limb? LIM homeodomain factors appear to regulate pathway selection through controlling expression of members of the Eph/ephrin signaling pathway. In the dorsally projecting lateral LMC population Lhx1 induces expression of the guidance receptor EphA4, and the presence of this receptor repels these axons from ephrin-A expressing cells in the ventral limb mesenchyme (Kania and Jessell, 2003). Both *EphA4* and *Lhx1* mutants are characterized by defects in dorsoventral projection choice (Helmbacher *et al.*, 2000; Kania *et al.*, 2000), and in *Lhx1* mutants lateral LMC neurons appear to select the dorsal and ventral compartments in a stochastic manner (Fig. 4.3B). Interestingly, in the limb mesenchyme another LIM homeodomain protein, Lmx1b, controls the expression of ephrin-A proteins in the ventral compartment (Kania *et al.*, 2000). Thus, these receptor–ligand interactions appear to be coordinated through actions of LIM homeodomain proteins expressed in neurons and limb mesenchyme.

Recently several additional factors have been implicated in the dorsoventral selection of LMC motor axons including members of the semaphorins/plexin and GDNF/ret signaling pathways (Huber *et al.*, 2005; Kramer *et al.*, 2006a). In mice lacking GDNF/ret signaling, dorsally projecting LMC axons follow an aberrant ventral trajectory, suggesting that this system functions as an instructive guidance signal for LMC neurons (Kramer *et al.*, 2006a). In mice mutant for members of the semaphorin/plexin family, medial and lateral LMC motor axons also display dorsoventral guidance defects, and this phenotype is associated with the defasiculation of axons as they project into the limb. The details of how these signaling pathways interact with one another in the control motor axon guidance decisions are still unclear, and are confounded by the observation that they operate within multiple phases of the limb innervation program. In addition, whether the LIM homeodomain code for dorsoventral trajectories regulates the expression of members of the semaphoring/plexin and GDNF/ret families remains to be determined.

A similar LIM homeodomain-based strategy appears to operate in the regulation of the expression of guidance receptors that control motor neuron innervation of axial musculature. Here the activity of the LIM homeodomain factor Lhx3, which becomes selectively expressed by axial muscle-innervating motor neurons, is sufficient to drive nearly all motor axons to axial muscle at the expense of other pathways (Sharma *et al.*, 2000). This action of Lhx3 appears to be mediated through the regulation of type I FGF receptor expression in motor neurons and FGF cues in the periphery (Shirasaki *et al.*, 2006). The use of LIM homeodomain proteins in motor neuron is conserved in diverse species such as flies and worms (Shirasaki and

Pfaff, 2002), suggesting that the actions of this family represent a highly conserved strategy in the early phases of motor axon guidance.

3.2. Controlling the fine specificity of motor neuron–muscle connectivity

While the regulation of dorsoventral axonal projections in the limb appears to be controlled through LIM homeodomain regulation of Eph receptor expression, significantly less is known about the mechanisms which guide motor axons to their muscle targets. After making their dorsoventral choice in the limb, motor axons follow highly stereotypic pathways along nerve trunks and branches that guide them to the position of newly formed muscle masses (Landmesser, 2001). As with dorsoventral choices, the projection of axons along the anterior–posterior and proximal–distal axes of the limb are aligned with cell body position within the spinal cord. In general, more rostrally positioned motor pools project toward anterior/proximal limb regions while more caudally positioned pools project into posterior/distal regions (Hollyday and Jacobson, 1990; Landmesser, 1978b). Some aspects to the control of this motor neuron–muscle topography can be linked to the actions of Hox-regulated pool-specific transcription factors.

Within LMC neurons, one critical output of Hox activity is to control the expression of a diverse array of motor pool-specific transcription factors, including members of the Nkx, Runx, POU, and Fox families (Cohen *et al.*, 2005; Dasen *et al.*, 2005; De Marco Garcia and Jessell, 2008; Dou *et al.*, 1997; Stifani *et al.*, 2008). A recent study provides evidence for a role for the Hox-dependent homeodomain transcription factor Nkx6.1 in the establishment of muscle-specific trajectories. While Nkx6 factors are initially expressed and necessary in all motor neuron progenitors (Vallstedt *et al.*, 2001), expression of Nkx6.1 becomes restricted to motor pools at limb levels at the time when motor axons project toward their target muscles. Expression of Nkx6.1 in pools appears to be programmed solely through motor neuron intrinsic transcriptional programs, as ablation of the limb has no affect on its expression in postmitotic neurons (De Marco Garcia and Jessell, 2008). In *Nkx6.1* mutant mice, each of the lumbar LMC pools that normally express Nkx6.1 exhibits early defects in muscle nerve branch formation, and a later reduction in muscle innervation (Fig. 4.3C) (De Marco Garcia and Jessell, 2008). The cell surface molecules which may be controlled by Nkx6.1 and involved in directing axons toward their specified target muscles remain to be identified.

Although the actions of pool-specific transcription factors appear to be required for the specificity of innervation of limb muscles, these factors alone are unlikely to account for all aspects of muscle nerve branch patterns and target selectivity. While several transcription factor families are expressed by motor neuron pools in a manner that is independent of limb-derived signals, only a handful of factors have been demonstrate to be expressed in specific pools. Whether there exists a single factor for each of the approximately 50 motor pool identities is uncertain. Alternatively, the combinatorial actions of Hox proteins could directly regulate expression of receptors that instruct muscle nerve trajectories without the involvement of intermediary transcription factors. Moreover, since aspects of motor pool identity rely on feedback control from the muscle itself, as well as patterned neural activity in the spinal cord, the cell intrinsic programs likely do not account for every aspect of the limb projection pattern.

3.3. Target-dependent steps in muscle nerve innervation patterns

While many aspects of the Hox/FoxP1 network in LMC neurons emerge in a target-independent manner, expression of certain pool-specific transcription factors relies on the presence signals from the periphery. Expression of the ETS transcription factors Pea3 and Er81 in motor pools depends on limb-derived neurotrophic signals provided by the limb mesoderm and muscle targets (Haase *et al.*, 2002; Lin *et al.*, 1998). These signals appear to be permissive rather than instructive and not all motor neurons are competent in their ability to respond to neurotrophic signals. One peripheral signal, glial-derived neurotrophic factor (GDNF) is expressed by the limb mesenchyme and muscle and appears to instruct the transcriptional identity of forelimb motor pools through regulation of Pea3 expression. In explants of spinal cord treated with GDNF, Pea3 is induced in a pattern approximating the normal number *in vivo* and is confined to the level of the spinal cord which normally expresses Pea3 (Haase *et al.*, 2002). Thus, not all motor neurons are equivalent in their ability to respond to GDNF, and this constrainment appears to be set by the profile of Hox expression in forelimb motor pools (Dasen *et al.*, 2005; Tiret *et al.*, 1998). Additional signals provided through Met signaling appear to enhance the influences of GDNF, through recruitment of additional neurons to the Pea3 motor pools (Helmbacher *et al.*, 2003).

These observations suggest that despite the critical role of Hox factors in motor neuron specification, target-derived cues also contribute to the transcriptional programming of pool fates. Expression of the target-induced factor Pea3 is critical for later aspects of motor pool differentiation such as the clustering of motor neurons into pools and muscle-specific patterns of axonal innervation (Livet *et al.*, 2002; Vrieseling and Arber, 2006). In *Pea3* mutants, motor axons are capable of reaching their appropriate muscle target, but fail to properly arborize the entire length of the muscle (Fig. 4.3D). Thus, motor pool specification appears to unfold in two main phases: an intrinsic phase that confers aspects of motor neuron identity involved in the selection of target muscle connectivity (Landmesser, 2001;

Milner and Landmesser, 1999), and a later phase that operates after motor axons have reached their muscle targets, which is associated with ETS gene expression and the clustering of motor neurons within the LMC (Livet *et al.*, 2002; Price *et al.*, 2002).

4. Control of Sensory Neuron Specification and Connectivity

A major sensory pathway from the body to the central nervous system is mediated through neurons whose cell bodies are located outside the spinal cord within the dorsal root ganglia (DRG). Like motor neurons, DRG sensory neurons are relatively well characterized in terms of their early specification programs and physiological functions (Chen *et al.*, 2003; Marmigere and Ernfors, 2007). However unlike motor neurons, sensory neurons are not organized into discrete columns and pools nor does the position of their cell bodies form topographic maps. In contrast, sensory neurons projecting to a given target appear to be scattered throughout the DRG (Fig. 4.4A). Nevertheless, our understanding of the genetic programs that contribute to the central and peripheral connectivity of sensory neurons has progressed significantly in recent years.

4.1. Early steps in sensory neuron lineage specification

As in the ventral spinal cord, the specification of different classes of sensory neurons appears to have its origins in a set of inductive signaling events operating along the dorsoventral axis of the neural tube. Sensory neurons residing within the DRG are derived from a migratory population of neural crest cells that are specified at the border between neural and non-neural ectoderm in the dorsal neural tube (Sauka-Spengler and Bronner-Fraser, 2008). Graded TGFβ and Wnt signaling from the roof plate and surface ectoderm appear to be essential in specifying the neural crest as a whole, as well as several classes of sensory relay neurons which populate the dorsal half of the spinal cord (Fig. 4.4A). The initial phases of neural crest specification are characterized by the delamination of cells from the neural tube and the expression of early transcription factors common to most neural crest progenitors, such as the HMG transcription factor Sox10 (Kim *et al.*, 2003).

The neural crest derivatives are a remarkably diverse class which, in addition to peripheral sensory ganglia, gives rise to the autonomic nervous system and several non-neuronal cell types (Marmigere and Ernfors, 2007; Sauka-Spengler and Bronner-Fraser, 2008). Recent studies indicate that the sensory lineages of the neural crest are specified in response to high levels of canonical Wnt signaling from the roof plate and surface ectoderm.

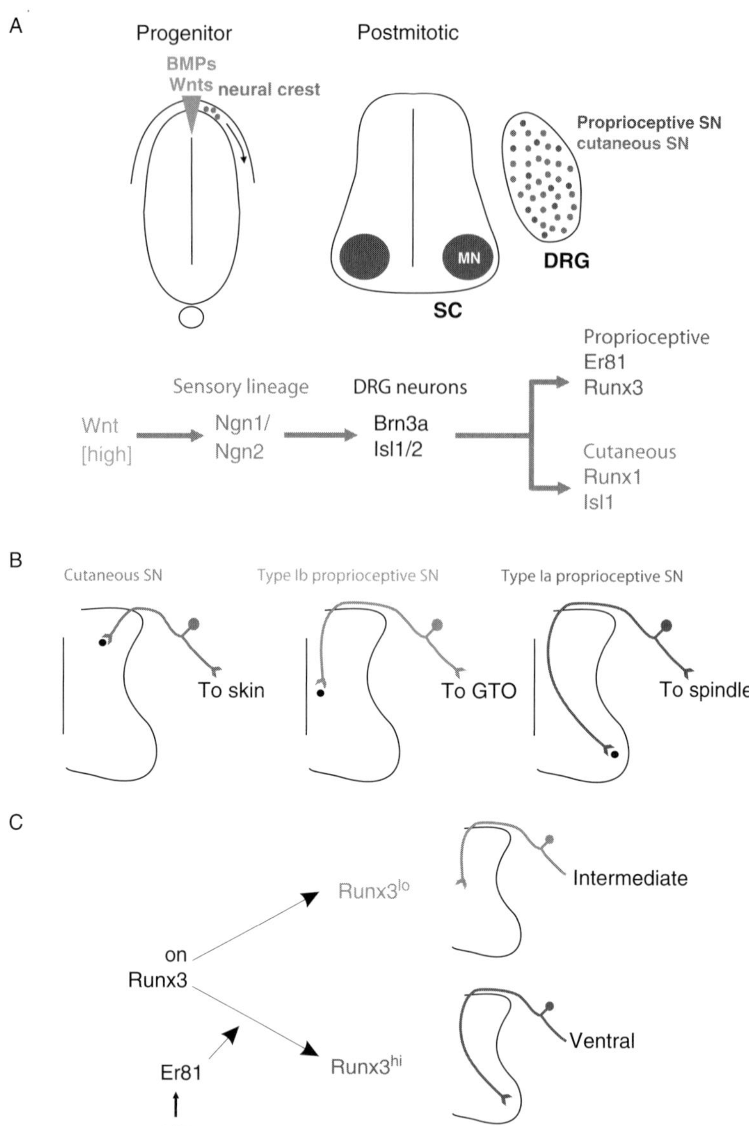

Figure 4.4 Specification of DRG sensory neuron identity and patterns of connectivity. (A) Neural crest cells are specified in response to TGFβ/BMP and Wnt signaling from the roof plate and surface ectoderm. Neural crest cells subsequently give rise to DRG sensory neurons, neurons in the autonomic nervous system, and multiple non-neuronal lineages. The early precursors to the sensory neural (SN) lineage are characterized by the expression of the bHLH factors Ngn1 and Ngn2, likely in response to high levels of canonical Wnt signaling. Postmitotic neurons populating the DRG express the POU domain protein Brn3a and the LIM homeodomain proteins Isl1 and Isl2. These populations segregate into two main classes: cutaneous sensory neurons which express and

Conditional mutation of *β-catenin* (a primary target of canonical Wnt signaling) in neural crest leads to the selective loss of sensory lineages and melanocytes, while other derivatives are relatively spared (Hari *et al.*, 2002). When neural crest cells are forced to express a constitutively active form of β-catenin, sensory neurons are generated at the expense of other neural crest lineages (Lee *et al.*, 2004). Thus, high levels of Wnt/β-catenin activity appear to exert an instructive role in the specification of sensory neurons.

As in the ventral spinal cord, the generation of diverse neural crest derivatives involves combinatorial transcription factor codes in progenitor and postmitotic neurons (Anderson, 1999). The sensory lineage requires the activities of two bHLH proteins, Neurogenin1 (Ngn1) and Neurogenin2 (Ngn2), as in *Ngn1/Ngn2* double-knockout mice the sensory lineages are selectively lost (Ma *et al.*, 1999). Multiple postmitotically expressed transcription factors common to all sensory lineages have also been characterized including the POU domain factor Brn3a, the LIM homeodomain factors Isl1 and Isl2, and the bHLH protein NeuroD (Sauka-Spengler and Bronner-Fraser, 2008). While each of these factors is expressed by all sensory lineages, their activities are often restricted to specific classes. For example, while Brn3a is required in all sensory neurons, Isl1 appears to exert a selective role in the diversification of sensory neurons that innervate the skin (Eng *et al.*, 2001; McEvilly *et al.*, 1996; Sun *et al.*, 2008).

Neurons within the sensory lineage further diversify into two main classes: proprioceptive neurons, which relay information about muscle tension and length, and cutaneous neurons, which relay several sensory modalities from the skin. These distinct sensory neuron classes can be anatomically defined by the termination of their axons within the spinal cord as well as by expression of receptors for neurotrophic factors (Marmigere and Ernfors, 2007). Cutaneous sensory neurons express the neurotrophin receptor TrkA and terminate within the dorsal spinal cord where they synapse with projection relay neurons. Proprioceptive neurons express TrkC, terminate in the intermediate and ventral spinal cord, and form connections with multiple ventral classes including local circuit interneurons, ascending projection neurons, and motor neurons.

require Runx1 and Isl1 function, and proprioceptive neurons that require Runx3 and Er81. (B) Different classes of sensory neurons can be characterized by their peripheral and central projection patterns. Cutaneous neurons project peripherally to skin and centrally to dorsal laminae of spinal cord (SC), type Ib proprioceptives project to Golgi tendon organs (GTOs) and project to the intermediate spinal cord, while type Ia project to muscle spindles and neuron in the ventral spinal cord. (C) Model for the control of central projections of proprioceptive afferents by Runx3. The level of Runx3 activity appears to control the ventral extent of proprioceptive afferents. Elevated levels of Runx3 appear to drive projections more ventrally. Runx3 activity levels may be controlled by interactions with Er81, which may be indirectly regulated through peripheral NT-3 signaling.

The genetic programs which differentiate cutaneous from proprioceptive neurons are not well understood. Both sensory classes are born in sequential manner, with the birth of proprioceptive neurons preceding generation of cutaneous sensory neurons. In addition to the early roles of Ngn1 and Ngn2 in specifying all sensory lineages, these factors also appear to be critically involved in the sequential generation of each of the two major classes. In *Ngn1* mutants, there is a complete absence of TrkA+ cutaneous neurons while proprioceptive neurons are relatively spared (Ma *et al.*, 1999). While in *Ngn2* mutants the early specification of proprioceptive neurons is selectively affected, *Ngn1* appears to be capable of compensating for loss of *Ngn2* in later stages.

Cutaneous and proprioceptive classes are also distinguished from one another by differential expression of two runt-domain transcription factors: Runx1 and Runx3. Runx1 is expressed in all TrkA+ sensory neurons and is required within a subpopulation of these cells for the expression of a variety of genes required for nociceptive (pain) sensation (Chen *et al.*, 2006b). Loss of *Runx1* leads to changes in the normal pattern of target innervation of nociceptive afferents in the dorsal horn, and the subtypes normally dependent on Runx1 function project to more dorsal regions of the spinal cord (Chen *et al.*, 2006b). In complimentary gain-of-function studies, it was also found that forced misexpression of Runx1 forces cutaneous neurons to project into deeper layers of the dorsal horn (Kramer *et al.*, 2006b). Similarly, *in vitro* studies indicate that Runx1 can promote axonal outgrowth in a dose-dependent manner (Marmigere *et al.*, 2006). As described below, Runx3 appears to exert a similar instructive role in the specification and axonal targeting of proprioceptive sensory neurons.

4.2. Genetic control of proprioceptive sensory neuron identity

Because proprioceptive neurons relay information about limb position and the contractile status of muscles to the CNS, their ability to form selective connections with motor neurons and interneurons is critical for coordinated locomotion. Like motor neurons, proprioceptive sensory neurons require a set of postmitotically expressed factors that are shared amongst most subtypes. One aspect of the transcriptional programming of their identities appears to drive the central projections of their axons toward the ventral spinal cord. This program has been shown to be mediated through the coordinated actions of two transcriptions factors, the ETS factor Er81 and Runx3.

In mice Er81 is expressed by virtually all TrkC+ proprioceptive neurons. In *Er81* mutants, the peripheral projections of sensory afferents to muscle appear to be established normally, and early aspects of muscle spindle formation are unaffected (Arber *et al.*, 2000). Centrally however, proprioceptive neurons fail to project to the ventral spinal cord, and the monosynaptic connections that form between proprioceptive neurons and motor neurons are not established. Electrophysiological studies confirm a loss of the monosynaptic stretch-reflex circuit and *Er81* mutant mice are characterized by severe defects in the coordination of limb movements. Thus, Er81 is required generally for the central connections between sensory neurons and their targets in the ventral spinal cord.

Within the proprioceptive class, sensory neurons further diversify into three major subtypes, termed group Ia, Ib, and II afferents, which differ in their peripheral connections and the extent of their ventral projections (Brown, 1981). Peripherally, group Ia and group II neurons project to muscle spindles while group Ib neurons project to Golgi tendon organs. Centrally, Group Ib afferents do not extend beyond the intermediate spinal cord, and group II afferents project only sparsely into the ventral spinal cord, while group Ia afferents send extensive ventral collateral projections (Fig. 4.4B). Because different laminar zones are populated by distinct classes of projection neurons (Brown, 1981), the selection of sensory axonal termination position has a critical role in establishing precise patterns of sensory connectivity.

How are the distinct termination zones of different proprioceptive afferent subtypes determined? The graded activities of the transcription factor Runx3 appear to exert a central role in the ventral extent of sensory neuron central projection. Like Er81, Runx3 is expressed by all proprioceptive subtypes, and in *Runx3* mutants proprioceptive neurons fail to innervate the ventral spinal cord (Inoue *et al.*, 2002). Gain-of-function studies indicate that Runx3 has a distinct role from Er81 in defining the central projection patterns of proprioceptive afferents. Within the proprioceptive population, the levels of Runx3 vary widely between individual neurons (Chen *et al.*, 2006a). Misexpression studies in the chick suggest that differences in the levels of Runx3 activity in proprioceptive neurons exert a profound influence on axonal projection patterns (Fig. 4.4C). Misexpression or overexpression of Runx3 in sensory neurons that terminate more dorsally can force these populations to project toward the ventral spinal cord. The ability of Runx3 to control the termination zone appears to be specific, as similar studies performed with Er81 overexpression had no effect on ventral projection patterns (Chen *et al.*, 2006a).

These findings suggest a model in which the different termination zones of group Ib, II, and Ia proprioceptive axons are specified by graded Runx3 activity. However, the mechanisms by which sensory neurons acquire different levels of Runx3 remains unclear. Different levels of Runx3 activity could be set by other factors expressed by proprioceptive sensory neurons which elevate the levels of Runx3 expression or modify its transcriptional activities. For example, Er81 could enhance the net concentration Runx3 in Ia afferents, bringing levels in many proprioceptive sensory neurons above a critical threshold that is needed to direct axons into the ventral spinal cord (Fig. 4.4C). Alternatively, Er81 may enhance Runx3 activity through direct protein interactions on target genes that are dependent on extrinsic or intrinsic cues.

As with motor neurons, aspects of the transcriptional identity of proprioceptive sensory neurons appear to be set by cues in the periphery. Expression of Er81 is controlled by peripheral NT-3 signaling (Patel *et al.*, 2003), and some aspects of the phenotypes observed in *Er81* mutants can be alleviated through over expression of NT-3 in muscle (Li *et al.*, 2006). How the NT-3/Er81 pathway influences the expression levels of Runx3 is not known, although one plausible model is that NT-3 also influences Runx3 expression which in turn may directly or indirectly control the laminar targeting of proprioceptive sensory neurons (Fig. 4.4C). As described below, the use of retrograde signaling by sensory neurons may extend beyond control of laminar targeting, and may define additional aspects of their innervation pattern, such as the synaptic specificity of connections between sensory neurons and motor neurons.

5. Sensory–Motor Circuit Assembly and Function

A relatively simple circuit in the nervous system is the monosynaptic stretch-reflex circuit which fundamentally consists of a motor neuron, a type Ia sensory afferent, and a muscle target (Fig. 4.5A) (Eccles *et al.*, 1957). When a muscle is stretched the activation of mechanoreceptors within muscle spindles leads to the excitation of Ia sensory afferents that synapse with motor neurons that innervate the same muscle. In addition to these direct monosynaptic inputs, proprioceptive neurons also form connections with inhibitory interneurons that are connected to motor neurons that project to functionally antagonistic muscles. This proprioceptive sensory feedback from muscle to neurons in the ventral spinal cord is essential in maintaining body posture and also has a critical role in adaptive responses to changes in the environment (Chen *et al.*, 2003; Dietz, 2002).

The central connections formed between sensory and motor neurons have been shown to be highly selective, as proprioceptive afferents projecting to a given muscle target avoid making monosynaptic connections with motor neurons innervating antagonistic or functionally unrelated muscles (Frank and Mendelson, 1990). Presumably, as Ia afferents reach the ventral spinal cord they must acquire the capacity to select only small subset of their many potential synaptic partners. Remarkably, it has been shown that a single group Ia sensory afferent generates several collateral branches that form synapses with the hundreds of motor neurons within a pool that innervate a homonymous muscle target (Mendell and Henneman, 1968).

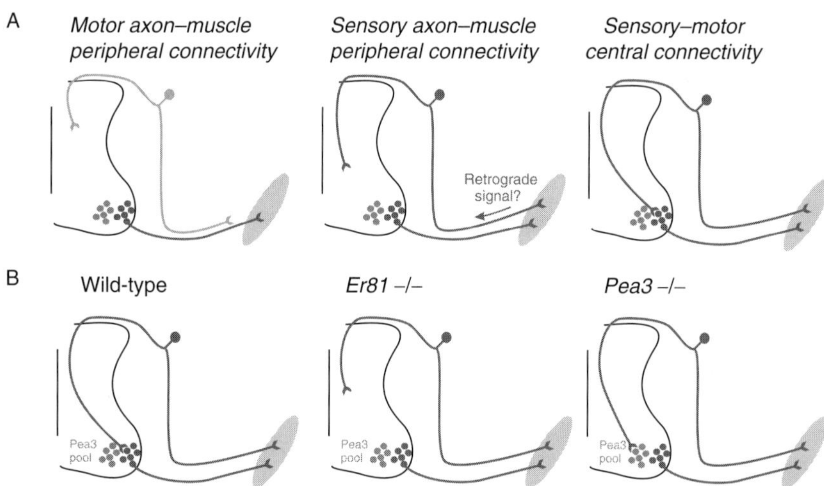

Figure 4.5 Models for the formation of the Ia monosynaptic stretch-reflex circuit. (A) In the initial phases of the formation of the stretch-reflex circuit, motor axons project to specific muscle targets by virtue of their Hox-induced transcriptional identity. In this model, sensory neurons are naïve with respect to their identities, and project along the pathways defined by motor axons. When sensory neurons reach their targets, a retrograde signal may allow them to acquire a muscle-specific identity. The acquisition of a muscle-specific identity by sensory neurons determines their capacity to selectively innervate motor neurons which project to the same muscle target. (B) Role of ETS transcription factors in sensory–motor connectivity. In *Er81* mutants, all proprioceptive sensory neurons fail to reach the ventral spinal cord. In *Pea3* mutants, loss of Pea3 activity in motor neurons leads to inappropriate innervation by other sensory neuron populations.

Because the specificity of connections appears to form in an activity-independent manner (Mendelson and Frank, 1991), the stretch-reflex circuit provides an attractive system to explore the genetic basis of synaptic specificity.

5.1. Early studies on the peripheral and central connectivity of proprioceptive neurons

One potential issue in trying to understand the genetic basis for circuit assembly is whether the principles that apply to one class of neurons are applicable to other classes. Does the concept of a transcription factor-based cell intrinsic program that defines the "muscle-specific" identity of motors neurons apply to the specification and connectivity patterns of sensory neurons? To what extent does the synaptic specificity between sensory neurons and motor neurons rely on cues from the periphery? Developmental studies in chick embryos have provided insights into the novel strategies used by sensory neurons in the formation of specific peripheral and central connections.

Since both motor and proprioceptive axons project to a variety muscle targets, one question is whether sensory neurons possess independent guidance mechanisms that confer the specificity of their peripheral projections. As motor axons project into the periphery, sensory axons appear to follow the routes pioneered by motor neurons (Gallarda *et al.*, 2008; Wang and Scott, 2000). Studies using chick embryological manipulations suggest that the peripheral projections of proprioceptive neurons are dependent on the routes established by motor axons. When motor neurons are surgically removed at a stage prior to when they project into the limb, sensory neurons are no longer capable of innervating muscle, but instead are diverted toward the targets cutaneous sensory neurons (Landmesser and Honig, 1986; Swanson and Lewis, 1986). A more recent study indicated that when motor neurons were ablated, sensory neurons were able to project to muscle, although some aspects in the specificity of these connections were eroded (Wang and Scott, 1999). These differences in the effects of motor neuron removal may reflect differences in the time period in which the manipulations were performed and hence the developmental potential of sensory neurons at different stages. Nevertheless, these studies suggest that unlike motor neurons, proprioceptive may be initially naïve, and their peripheral target specificities defined by the specificity of motor axon projections.

There is additional evidence that the specificity of the central connections of proprioceptive afferents within the spinal cord is acquired relatively late in development in response to signals provided by the periphery. Evidence in support of this view has come from analysis of the central innervation pattern of proprioceptive afferents in chick embryos grafted with a double-dorsal limb mesenchyme (Wenner and Frank, 1995). In this situation, the peripheral terminals of proprioceptive afferents are forced to innervate inappropriate limb muscles, and there is a corresponding change in central connectivity that matches the new peripheral target (Wenner and Frank, 1995). Aspects of sensory–motor target specificity may therefore be imparted by some specialized property of the muscle group itself, possibly through retrograde signaling to sensory neuron cell body.

These observations have suggested a model in which the identity of motor neurons establishes one fixed element in the stretch-reflex circuit, with a later peripheral influence imposing a matching identity on group Ia proprioceptive afferents, with the consequence that incoming afferent fibers are able to recognize distinct motor neuron pools and select appropriate partners for synapse formation (Fig. 4.5A). How this "molecular matching" program is established in unknown but could involve sets of homophilic recognition molecules or complimentary ligand–receptor pairs in pre- and postsynaptic cells.

5.2. Feedback control and molecular matching of sensory and motor neurons

Recent studies exploring the steps that specify motor and sensory neuron subtypes provide evidence that the expression of common transcription factors or their targets define the specificity of connections in the stretch-reflex circuit. Thus far, the most compelling studies have emerged from the analysis of the role of motor neuron identity in the assembly of this circuit. In addition to its role in motor axon arborization patterns, the ETS protein Pea3 also appears to exert a role in the formation of stretch-reflex circuits for specific muscles in the forelimb. In *Pea3* mutants, the motor neurons that have lost Pea3 expression receive input from inappropriate motor pools (Fig. 4.5B) (Vrieseling and Arber, 2006), suggesting that Pea3 is necessary to restrict inappropriate sensory neurons from forming monosynaptic connections. One additional aspect of Pea3 function in sensory–motor connectivity is to configure the pattern of motor neuron dendrites, as in *Pea3* mutants the normal pattern of dendritic arborization is altered. Whether other pool-specific transcription factors exert similar roles in the synaptic specificity in sensory–motor circuits remains to be explored.

What are the possible molecular programs that might control the assembly of the monosynaptic stretch-reflex circuit? Expression of Pea3 in motor neurons is controlled by Hox proteins (Dasen *et al.*, 2005), and Pea3 is necessary for the expression of multiple surface recognition and guidance molecules such as type II cadherins and semaphorins (Livet *et al.*, 2002). Combinatorial expression of semaphorin expression has been suggested to be involved in motor axon guidance decisions in the limb (Cohen *et al.*, 2005), and several type II cadherin family members have been shown to be expressed in both sensory and motor neurons in a limb-dependent manner (Price *et al.*, 2002). In addition, certain *Hox* genes have been shown to be expressed by DRG sensory neurons in a pattern that parallels *Hox* patterns within the spinal cord (Belting *et al.*, 1998), consistent with the hypothesis they may be involved in the matching of sensory and motor neurons. These disparate observations raise the possibility that coordinate expression of *Hox* genes acts in the initial phases of motor and sensory diversification, while intermediate transcription factors, target-dependent programs, and synaptic specificity determinants acting to control the selectivity of sensory–motor connectivity.

Could the concept of a transcription factor-based molecular matching system explain how neural circuits are assembled in other regions of the nervous system? In the developing hindbrain, the *Hoxa2* gene has been shown to regulate the connections formed between motor neurons and sensory neurons and contribute to formation of facial somatotopic maps (Oury *et al.*, 2006). Similarly, the homeodomain protein Phox2b has been shown to function in different classes of neurons involved in the assembly of autonomic reflex circuits (Dauger *et al.*, 2003). However, the cell-type-specific functions of these genes have not been assessed, nor are the relevant downstream effectors of these transcription factors known. Nevertheless, these observations are consistent with the idea that aspects of synaptic specificity could be determined through expression of common transcription factors within diverse neuronal classes.

6. Conclusions

While our understanding of the early specification programs that control the synaptic specificity of motor and sensory neurons has progressed significantly, the extent to which they provide any insights into the genetic basis of innate motor behaviors remains to be seen. A recent study in *Drosophila* indicates that the activities of individual *Hox* genes can switch the pattern of motor output within embryonic segments and lead to homeotic transformation of larval motility behaviors (Dixit *et al.*, 2008). These observations are consistent with the idea that the activities of single genes, acting in the context of diverse neuronal classes, can reprogram stereotypic patterns of movement. It will be interesting to determine whether these influences of *Hox* genes in the *Drosophila* nervous system are mediated through their activities in motor neurons, sensory neurons, or interneurons, or perhaps all three classes.

Studies of the spinal locomotor system indicate that much the diversification of its resident neuronal classes and subtypes can be linked to the actions of a coherent set of transcription factors expressed in discrete domains along the dorsoventral and rostrocaudal axes. Studies in motor neurons indicate that much of this diversity is mediated through the actions of a single large family of transcription factors encoded by the *Hox* gene clusters. Whether the actions of these factors also influence the specificity of connections between motor neurons, interneurons, and sensory neurons, and contribute to locomotor behaviors, should be an area of exciting investigation in the future.

The control of movement relies on the integration of circuits residing in multiple regions of the nervous system outside the spinal cord, including the hindbrain, cerebellum, basal ganglia, and somatosensory cortex. A more complete understanding of the neural circuitries that specify innate motor behaviors will require a better picture of how these circuits are assembled during development, and how these regions integrate with spinal networks. Recent studies on the developmental programs that specify circuits in the cortex have revealed that pyramidal neurons and interneurons use very similar molecular strategies to drive neuronal differentiation, involving transcriptional networks active in progenitor and postmitotic cells (Fishell,

2007; Molyneaux *et al.*, 2007). Although cortical areas lack expression of the chromosomally arrayed *Hox* genes, these circuits may also use coherent families of transcription factors to control the matching of diverse neuronal classes during synaptogenesis.

ACKNOWLEDGMENTS

I would like to thank Tom Jessell for many discussions relating to the studies described in this chapter, and Molly Cahill for critical reading of the text. Work in my lab is supported by grants from the Burroughs Welcome Fund, the Alfred P. Sloan Foundation, and McKnight Foundation.

REFERENCES

Anderson, D. J. (1999). Lineages and transcription factors in the specification of vertebrate primary sensory neurons. *Curr. Opin. Neurobiol.* **9,** 517–524.

Arber, S., Han, B., Mendelsohn, M., Smith, M., Jessell, T. M., and Sockanathan, S. (1999). Requirement for the homeobox gene Hb9 in the consolidation of motor neuron identity. *Neuron* **23,** 659–674.

Arber, S., Ladle, D. R., Lin, J. H., Frank, E., and Jessell, T. M. (2000). ETS gene Er81 controls the formation of functional connections between group Ia sensory afferents and motor neurons. *Cell* **101,** 485–498.

Belting, H. G., Shashikant, C. S., and Ruddle, F. H. (1998). Multiple phases of expression and regulation of mouse Hoxc8 during early embryogenesis. *J. Exp. Zool.* **282,** 196–222.

Bel-Vialar, S., Itasaki, N., and Krumlauf, R. (2002). Initiating Hox gene expression: In the early chick neural tube differential sensitivity to FGF and RA signaling subdivides the HoxB genes in two distinct groups. *Development (Cambridge)* **129,** 5103–5115.

Briscoe, J., Pierani, A., Jessell, T. M., and Ericson, J. (2000). A homeodomain protein code specifies progenitor cell identity and neuronal fate in the ventral neural tube. *Cell* **101,** 435–445.

Brown, A. G. (1981). "Organization in the Spinal Cord: The Anatomy and Physiology of Identified Neurones." Springer-Verlag, Berlin.

Chen, A. I., de Nooij, J. C., and Jessell, T. M. (2006a). Graded activity of transcription factor Runx3 specifies the laminar termination pattern of sensory axons in the developing spinal cord. *Neuron* **49,** 395–408.

Chen, C. L., Broom, D. C., Liu, Y., de Nooij, J. C., Li, Z., Cen, C., Samad, O. A., Jessell, T. M., Woolf, C. J., and Ma, Q. (2006b). Runx1 determines nociceptive sensory neuron phenotype and is required for thermal and neuropathic pain. *Neuron* **49,** 365–377.

Chen, H. H., Hippenmeyer, S., Arber, S., and Frank, E. (2003). Development of the monosynaptic stretch reflex circuit. *Curr. Opin. Neurobiol.* **13,** 96–102.

Cohen, S., Funkelstein, L., Livet, J., Rougon, G., Henderson, C. E., Castellani, V., and Mann, F. (2005). A semaphorin code defines subpopulations of spinal motor neurons during mouse development. *Eur. J. Neurosci.* **21,** 1767–1776.

Dasen, J. S., Liu, J. P., and Jessell, T. M. (2003). Motor neuron columnar fate imposed by sequential phases of Hox-c activity. *Nature* **425,** 926–933.

Dasen, J. S., Tice, B. C., Brenner-Morton, S., and Jessell, T. M. (2005). A Hox regulatory network establishes motor neuron pool identity and target-muscle connectivity. *Cell* **123,** 477–491.

Dasen, J. S., De Camilli, A., Wang, B., Tucker, P. W., and Jessell, T. M. (2008). Hox repertoires for motor neuron diversity and connectivity gated by a single accessory factor, FoxP1. *Cell* **134,** 304–316.

Dauger, S., Pattyn, A., Lofaso, F., Gaultier, C., Goridis, C., Gallego, J., and Brunet, J. F. (2003). Phox2b controls the development of peripheral chemoreceptors and afferent visceral pathways. *Development (Cambridge)* **130,** 6635–6642.

De Marco Garcia, N. V., and Jessell, T. M. (2008). Early motor neuron pool identity and muscle nerve trajectory defined by postmitotic restrictions in Nkx6.1 activity. *Neuron* **57,** 217–231.

Demir, E., and Dickson, B. J. (2005). Fruitless splicing specifies male courtship behavior in *Drosophila*. *Cell* **121,** 785–794.

Dessaud, E., McMahon, A. P., and Briscoe, J. (2008). Pattern formation in the vertebrate neural tube: A sonic hedgehog morphogen-regulated transcriptional network. *Development (Cambridge)* **135,** 2489–2503.

Dietz, V. (2002). Proprioception and locomotor disorders. *Nat. Rev. Neurosci.* **3,** 781–790.

Dixit, R., Vijayraghavan, K., and Bate, M. (2008). Hox genes and the regulation of movement in *Drosophila*. *Dev. Neurobiol.* **68,** 309–316.

Dou, C., Ye, X., Stewart, C., Lai, E., and Li, S. C. (1997). TWH regulates the development of subsets of spinal cord neurons. *Neuron* **18,** 539–551.

Eccles, J. C., Eccles, R. M., and Lundberg, A. (1957). The convergence of monosynaptic excitatory afferents on to many different species of alpha motoneurones. *J. Physiol.* **137,** 22–50.

Eng, S. R., Gratwick, K., Rhee, J. M., Fedtsova, N., Gan, L., and Turner, E. E. (2001). Defects in sensory axon growth precede neuronal death in Brn3a-deficient mice. *J. Neurosci.* **21,** 541–549.

Ensini, M., Tsuchida, T. N., Belting, H. G., and Jessell, T. M. (1998). The control of rostrocaudal pattern in the developing spinal cord: Specification of motor neuron subtype identity is initiated by signals from paraxial mesoderm. *Development (Cambridge)* **125,** 969–982.

Fishell, G. (2007). Perspectives on the developmental origins of cortical interneuron diversity. *Novartis Found. Symp.* **288,** 21–35; discussion 35–44, 96–28.

Frank, E., and Mendelson, B. (1990). Specification of synaptic connections between sensory and motor neurons in the developing spinal cord. *J. Neurobiol.* **21,** 33–50.

Gallarda, B. W., Bonanomi, D., Muller, D., Brown, A., Alaynick, W. A., Andrews, S. E., Lemke, G., Pfaff, S. L., and Marquardt, T. (2008). Segregation of axial motor and sensory pathways via heterotypic trans-axonal signaling. *Science (New York)* **320,** 233–236.

Goulding, M., and Pfaff, S. L. (2005). Development of circuits that generate simple rhythmic behaviors in vertebrates. *Curr. Opin. Neurobiol.* **15,** 14–20.

Greene, E. C. (1963). "Anatomy of the Rat." Hafner Publishing Company, New York.

Haase, G., Dessaud, E., Garces, A., de Bovis, B., Birling, M., Filippi, P., Schmalbruch, H., Arber, S., and deLapeyriere, O. (2002). GDNF acts through PEA3 to regulate cell body positioning and muscle innervation of specific motor neuron pools. *Neuron* **35,** 893–905.

Hanson, M. G., Milner, L. D., and Landmesser, L. T. (2008). Spontaneous rhythmic activity in early chick spinal cord influences distinct motor axon pathfinding decisions. *Brain Res. Rev.* **57,** 77–85.

Hari, L., Brault, V., Kleber, M., Lee, H. Y., Ille, F., Leimeroth, R., Paratore, C., Suter, U., Kemler, R., and Sommer, L. (2002). Lineage-specific requirements of beta-catenin in neural crest development. *J. Cell Biol.* **159,** 867–880.

Helmbacher, F., Schneider-Maunoury, S., Topilko, P., Tiret, L., and Charnay, P. (2000). Targeting of the EphA4 tyrosine kinase receptor affects dorsal/ventral pathfinding of limb motor axons. *Development (Cambridge)* **127,** 3313–3324.

Helmbacher, F., Dessaud, E., Arber, S., deLapeyriere, O., Henderson, C. E., Klein, R., and Maina, F. (2003). Met signaling is required for recruitment of motor neurons to PEA3-positive motor pools. *Neuron* **39,** 767–777.

Hobert, O. (2003). Behavioral plasticity in *C. elegans*: Paradigms, circuits, genes. *J. Neurobiol.* **54,** 203–223.

Hollyday, M., and Jacobson, R. D. (1990). Location of motor pools innervating chick wing. *J. Comp. Neurol.* **302,** 575–588.

Huber, A. B., Kania, A., Tran, T. S., Gu, C., De Marco Garcia, N., Lieberam, I., Johnson, D., Jessell, T. M., Ginty, D. D., and Kolodkin, A. L. (2005). Distinct roles for secreted semaphorin signaling in spinal motor axon guidance. *Neuron* **48,** 949–964.

Inoue, K., Ozaki, S., Shiga, T., Ito, K., Masuda, T., Okado, N., Iseda, T., Kawaguchi, S., Ogawa, M., Bae, S. C., *et al.* (2002). Runx3 controls the axonal projection of proprioceptive dorsal root ganglion neurons. *Nat. Neurosci.* **5,** 946–954.

Jessell, T. M. (2000). Neuronal specification in the spinal cord: Inductive signals and transcriptional codes. *Nat. Rev. Genet.* **1,** 20–29.

Kania, A., and Jessell, T. M. (2003). Topographic motor projections in the limb imposed by LIM homeodomain protein regulation of ephrin-A:EphA interactions. *Neuron* **38,** 581–596.

Kania, A., Johnson, R. L., and Jessell, T. M. (2000). Coordinate roles for LIM homeobox genes in directing the dorsoventral trajectory of motor axons in the vertebrate limb. *Cell* **102,** 161–173.

Kiehn, O., and Butt, S. J. (2003). Physiological, anatomical and genetic identification of CPG neurons in the developing mammalian spinal cord. *Prog. Neurobiol.* **70,** 347–361.

Kim, J., Lo, L., Dormand, E., and Anderson, D. J. (2003). SOX10 maintains multipotency and inhibits neuronal differentiation of neural crest stem cells. *Neuron* **38,** 17–31.

Kmita, M., and Duboule, D. (2003). Organizing axes in time and space; 25 years of colinear tinkering. *Science (New York)* **301,** 331–333.

Kramer, E. R., Knott, L., Su, F., Dessaud, E., Krull, C. E., Helmbacher, F., and Klein, R. (2006a). Cooperation between GDNF/Ret and ephrinA/EphA4 signals for motor-axon pathway selection in the limb. *Neuron* **50,** 35–47.

Kramer, I., Sigrist, M., de Nooij, J. C., Taniuchi, I., Jessell, T. M., and Arber, S. (2006b). A role for Runx transcription factor signaling in dorsal root ganglion sensory neuron diversification. *Neuron* **49,** 379–393.

Ladle, D. R., Pecho-Vrieseling, E., and Arber, S. (2007). Assembly of motor circuits in the spinal cord: Driven to function by genetic and experience-dependent mechanisms. *Neuron* **56,** 270–283.

Landmesser, L. (1978a). The development of motor projection patterns in the chick hind limb. *J. Physiol.* **284,** 391–414.

Landmesser, L. (1978b). The distribution of motoneurones supplying chick hind limb muscles. *J. Physiol.* **284,** 371–389.

Landmesser, L., and Honig, M. G. (1986). Altered sensory projections in the chick hind limb following the early removal of motoneurons. *Dev. Biol.* **118,** 511–531.

Landmesser, L. T. (2001). The acquisition of motoneuron subtype identity and motor circuit formation. *Int. J. Dev. Neurosci.* **19,** 175–182.

Lanuza, G. M., Gosgnach, S., Pierani, A., Jessell, T. M., and Goulding, M. (2004). Genetic identification of spinal interneurons that coordinate left–right locomotor activity necessary for walking movements. *Neuron* **42,** 375–386.

Lee, H. Y., Kleber, M., Hari, L., Brault, V., Suter, U., Taketo, M. M., Kemler, R., and Sommer, L. (2004). Instructive role of Wnt/beta-catenin in sensory fate specification in neural crest stem cells. *Science (New York)* **303,** 1020–1023.

Li, L. Y., Wang, Z., Sedy, J., Quazi, R., Walro, J. M., Frank, E., and Kucera, J. (2006). Neurotrophin-3 ameliorates sensory–motor deficits in Er81-deficient mice. *Dev. Dyn.* **235,** 3039–3050.

Lin, J. H., Saito, T., Anderson, D. J., Lance-Jones, C., Jessell, T. M., and Arber, S. (1998). Functionally related motor neuron pool and muscle sensory afferent subtypes defined by coordinate ETS gene expression. *Cell* **95,** 393–407.

Liu, J. P., Laufer, E., and Jessell, T. M. (2001). Assigning the positional identity of spinal motor neurons: Rostrocaudal patterning of Hox-c expression by FGFs, Gdf11, and retinoids. *Neuron* **32,** 997–1012.

Livet, J., Sigrist, M., Stroebel, S., De Paola, V., Price, S. R., Henderson, C. E., Jessell, T. M., and Arber, S. (2002). ETS gene Pea3 controls the central position and terminal arborization of specific motor neuron pools. *Neuron* **35,** 877–892.

Ma, Q., Fode, C., Guillemot, F., and Anderson, D. J. (1999). Neurogenin1 and neurogenin2 control two distinct waves of neurogenesis in developing dorsal root ganglia. *Genes Dev.* **13,** 1717–1728.

Marmigere, F., and Ernfors, P. (2007). Specification and connectivity of neuronal subtypes in the sensory lineage. *Nat. Rev. Neurosci.* **8,** 114–127.

Marmigere, F., Montelius, A., Wegner, M., Groner, Y., Reichardt, L. F., and Ernfors, P. (2006). The Runx1/AML1 transcription factor selectively regulates development and survival of TrkA nociceptive sensory neurons. *Nat. Neurosci.* **9,** 180–187.

McEvilly, R. J., Erkman, L., Luo, L., Sawchenko, P. E., Ryan, A. F., and Rosenfeld, M. G. (1996). Requirement for Brn-3.0 in differentiation and survival of sensory and motor neurons. *Nature* **384,** 574–577.

Mendell, L. M., and Henneman, E. (1968). Terminals of single Ia fibers: Distribution within a pool of 300 homonymous motor neurons. *Science (New York)* **160,** 96–98.

Mendelson, B., and Frank, E. (1991). Specific monosynaptic sensory–motor connections form in the absence of patterned neural activity and motoneuronal cell death. *J. Neurosci.* **11,** 1390–1403.

Milner, L. D., and Landmesser, L. T. (1999). Cholinergic and GABAergic inputs drive patterned spontaneous motoneuron activity before target contact. *J. Neurosci.* **19,** 3007–3022.

Molyneaux, B. J., Arlotta, P., and Macklis, J. D. (2007). Molecular development of corticospinal motor neuron circuitry. *Novartis Found. Symp.* **288,** 3–15; discussion 15–20, 96–18.

Muhr, J., Andersson, E., Persson, M., Jessell, T. M., and Ericson, J. (2001). Groucho-mediated transcriptional repression establishes progenitor cell pattern and neuronal fate in the ventral neural tube. *Cell* **104,** 861–873.

Novitch, B. G., Chen, A. I., and Jessell, T. M. (2001). Coordinate regulation of motor neuron subtype identity and pan-neuronal properties by the bHLH repressor Olig2. *Neuron* **31,** 773–789.

Oury, F., Murakami, Y., Renaud, J. S., Pasqualetti, M., Charnay, P., Ren, S. Y., and Rijli, F. M. (2006). Hoxa2- and rhombomere-dependent development of the mouse facial somatosensory map. *Science (New York)* **313,** 1408–1413.

Patel, T. D., Kramer, I., Kucera, J., Niederkofler, V., Jessell, T. M., Arber, S., and Snider, W. D. (2003). Peripheral NT3 signaling is required for ETS protein expression and central patterning of proprioceptive sensory afferents. *Neuron* **38,** 403–416.

Price, S. R., De Marco Garcia, N. V., Ranscht, B., and Jessell, T. M. (2002). Regulation of motor neuron pool sorting by differential expression of type II cadherins. *Cell* **109,** 205–216.

Rousso, D. L., Gaber, Z. B., Wellik, D., Morrisey, E. E., and Novitch, B. G. (2008). Coordinated actions of the forkhead protein Foxp1 and Hox proteins in the columnar organization of spinal motor neurons. *Neuron* **59,** 226–240.

Sauka-Spengler, T., and Bronner-Fraser, M. (2008). A gene regulatory network orchestrates neural crest formation. *Nat. Rev.* **9,** 557–568.

Shah, V., Drill, E., and Lance-Jones, C. (2004). Ectopic expression of Hoxd10 in thoracic spinal segments induces motoneurons with a lumbosacral molecular profile and axon projections to the limb. *Dev. Dyn.* **231,** 43–56.

Sharma, K., and Belmonte, J. C. (2001). Development of the limb neuromuscular system. *Curr. Opin. Cell Biol.* **13,** 204–210.

Sharma, K., Leonard, A. E., Lettieri, K., and Pfaff, S. L. (2000). Genetic and epigenetic mechanisms contribute to motor neuron pathfinding. *Nature* **406,** 515–519.

Shirasaki, R., and Pfaff, S. L. (2002). Transcriptional codes and the control of neuronal identity. *Annu. Rev. Neurosci.* **25,** 251–281.

Shirasaki, R., Lewcock, J. W., Lettieri, K., and Pfaff, S. L. (2006). FGF as a target-derived chemoattractant for developing motor axons genetically programmed by the LIM code. *Neuron* **50,** 841–853.

Sockanathan, S., and Jessell, T. M. (1998). Motor neuron-derived retinoid signaling specifies the subtype identity of spinal motor neurons. *Cell* **94,** 503–514.

Stifani, N., Freitas, A. R., Liakhovitskaia, A., Medvinsky, A., Kania, A., and Stifani, S. (2008). Suppression of interneuron programs and maintenance of selected spinal motor neuron fates by the transcription factor AML1/Runx1. *Proc. Natl. Acad. Sci. USA* **105,** 6451–6456.

Sullivan, G. E. (1962). Anatomy and embryology of the wing musculature of the domestic fowl (Gallus). *Aust. J. Zool.* **10,** 458–518.

Sun, Y., Dykes, I. M., Liang, X., Eng, S. R., Evans, S. M., and Turner, E. E. (2008). A central role for Islet1 in sensory neuron development linking sensory and spinal gene regulatory programs. *Nat. Neurosci.* **11,** 1283–1293.

Swanson, G. J., and Lewis, J. (1986). Sensory nerve routes in chick wing buds deprived of motor innervation. *J. Embryol. Exp. Morphol.* **95,** 37–52.

Tanabe, Y., William, C., and Jessell, T. M. (1998). Specification of motor neuron identity by the MNR2 homeodomain protein. *Cell* **95,** 67–80.

Tarchini, B., Huynh, T. H., Cox, G. A., and Duboule, D. (2005). HoxD cluster scanning deletions identify multiple defects leading to paralysis in the mouse mutant Ironside. *Genes Dev.* **19,** 2862–2876.

Thaler, J., Harrison, K., Sharma, K., Lettieri, K., Kehrl, J., and Pfaff, S. L. (1999). Active suppression of interneuron programs within developing motor neurons revealed by analysis of homeodomain factor HB9. *Neuron* **23,** 675–687.

Thaler, J. P., Koo, S. J., Kania, A., Lettieri, K., Andrews, S., Cox, C., Jessell, T. M., and Pfaff, S. L. (2004). A postmitotic role for Isl-class LIM homeodomain proteins in the assignment of visceral spinal motor neuron identity. *Neuron* **41,** 337–350.

Tiret, L., Le Mouellic, H., Maury, M., and Brulet, P. (1998). Increased apoptosis of motoneurons and altered somatotopic maps in the brachial spinal cord of Hoxc-8-deficient mice. *Development (Cambridge)* **125,** 279–291.

Tosney, K. W., and Landmesser, L. T. (1985a). Development of the major pathways for neurite outgrowth in the chick hindlimb. *Dev. Biol.* **109,** 193–214.

Tosney, K. W., and Landmesser, L. T. (1985b). Growth cone morphology and trajectory in the lumbosacral region of the chick embryo. *J. Neurosci.* **5,** 2345–2358.

Tsuchida, T., Ensini, M., Morton, S. B., Baldassare, M., Edlund, T., Jessell, T. M., and Pfaff, S. L. (1994). Topographic organization of embryonic motor neurons defined by expression of LIM homeobox genes. *Cell* **79,** 957–970.

Vallstedt, A., Muhr, J., Pattyn, A., Pierani, A., Mendelsohn, M., Sander, M., Jessell, T. M., and Ericson, J. (2001). Different levels of repressor activity assign redundant and specific roles to Nkx6 genes in motor neuron and interneuron specification. *Neuron* **31,** 743–755.

Vrieseling, E., and Arber, S. (2006). Target-induced transcriptional control of dendritic patterning and connectivity in motor neurons by the ETS gene Pea3. *Cell* **127,** 1439–1452.

Wang, G., and Scott, S. A. (1999). Independent development of sensory and motor innervation patterns in embryonic chick hindlimbs. *Dev. Biol.* **208,** 324–336.

Wang, G., and Scott, S. A. (2000). The "waiting period" of sensory and motor axons in early chick hindlimb: Its role in axon pathfinding and neuronal maturation. *J. Neurosci.* **20,** 5358–5366.

Wenner, P., and Frank, E. (1995). Peripheral target specification of synaptic connectivity of muscle spindle sensory neurons with spinal motoneurons. *J. Neurosci.* **15,** 8191–8198.

Wu, Y., Wang, G., Scott, S. A., and Capecchi, M. R. (2008). Hoxc10 and Hoxd10 regulate mouse columnar, divisional and motor pool identity of lumbar motoneurons. *Development (Cambridge)* **135,** 171–182.

CHAPTER FIVE

Development of Neural Circuits in the Adult Hippocampus

Yan Li,[1] Yangling Mu,[1] *and* Fred H. Gage

Contents

Abstract

The hippocampal formation is widely studied in part because of its distinct and highly laminar organization as well as its demonstrated fundamental role in learning and memory. The dentate gyrus of the hippocampal formation is one of two recognized brain regions that continually generate new neurons in adulthood. In this chapter, we review the basic structure of hippocampal cellular components and circuitry, the properties of stem cells and their progeny in the dentate gyrus, and the known mechanisms and timing of their maturation and integration into the adult circuitry. We also address the functional implication of neurogenesis in the adult hippocampus.

Laboratory of Genetics LOG-G, The Salk Institute for Biological Studies, La Jolla, California, USA
[1] These authors contributed equally to this work

Current Topics in Developmental Biology, Volume 87
ISSN 0070-2153, DOI: 10.1016/S0070-2153(09)01205-8

1. Development of Neural Circuits in the Adult Hippocampus

The hippocampus is one of the most studied areas of the mammalian central nervous system and has been shown to play a fundamental role in some forms of learning and memory (Brown *et al.*, 1990; Squire, 1987). The hippocampus is part of a functional brain system called the hippocampal formation, which includes the dentate gyrus, hippocampus, subiculum, presubiculum, parasubiculum, and entorhinal cortex. The hippocampus has a distinctive and easily identifiable neuroanatomy; the basic circuitry of the hippocampal formation has been known since the time of Ramon y Cajal, 100 years ago. The unidirectional progression of the excitatory pathway links each region of the hippocampal formation and forms a trisynaptic circuit in the hippocampus (Fig. 5.1) (Andersen *et al.*, 1966a,b).

Most sensory information reaches the hippocampus through the entorhinal cortex, which is considered to be the starting point of the trisynaptic hippocampal circuit (Swanson, 1983). Neurons in layer II of the entorhinal cortex give rise to axons that project through the subiculum to the dentate gyrus and CA3 region of hippocampus (Steward, 1976). The projections from the entorhinal cortex to the dentate gyrus form part of the major hippocampal input pathway, called the perforant pathway. The projection from the medial entorhinal cortex terminates within the middle portion of the molecular layer of the dentate gyrus, while that from the lateral entorhinal cortex terminates in the outer third of the molecular layer. These two

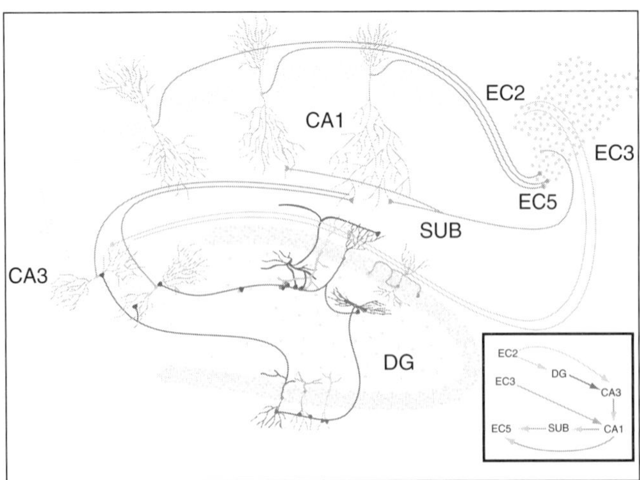

Figure 5.1 The hippocampal network.

parts of the perforant pathway also terminate in a laminar pattern in the stratum lacunosum-moleculare of CA3 and CA2. Neurons in layer III of the entorhinal cortex project to CA1 and the subiculum. Projections arising from the medial entorhinal cortex terminate in the stratum lacunosum-moleculare of CA1, which is close to CA3; and those from the lateral entorhinal cortex end in the portion of the stratum lacunosum-moleculare that is at the border between CA1 and the subiculum.

The dentate gyrus is the next step in the trisynaptic connection (Fig. 5.2). Its principal neurons, the dentate granule cells (DGCs), give rise to axons called mossy fibers that terminate on the proximal dendrites of CA3 pyramidal cells (Claiborne *et al.*, 1986). The mossy fibers also contact mossy cells and interneurons in the hilus. The CA3 pyramidal cells project heavily to other levels of the CA3, as well as to CA1. The projection from CA3 to CA1 is called the Schaffer collateral projection (Andersen, 1975). CA1 pyramidal cells project to both the subiculum and deep layers of the entorhinal cortex (Fig. 5.1). The deep layers of the entorhinal cortex, in turn, give rise to projection back to many of the same cortical areas that originally projected to the entorhinal cortex (Kohler, 1986). Thus, sensory information converging on the entorhinal cortex from the specific cortical areas by way of those excitatory pathways described above proceeds through the hippocampal circuit and then returns to the cortical region of origin.

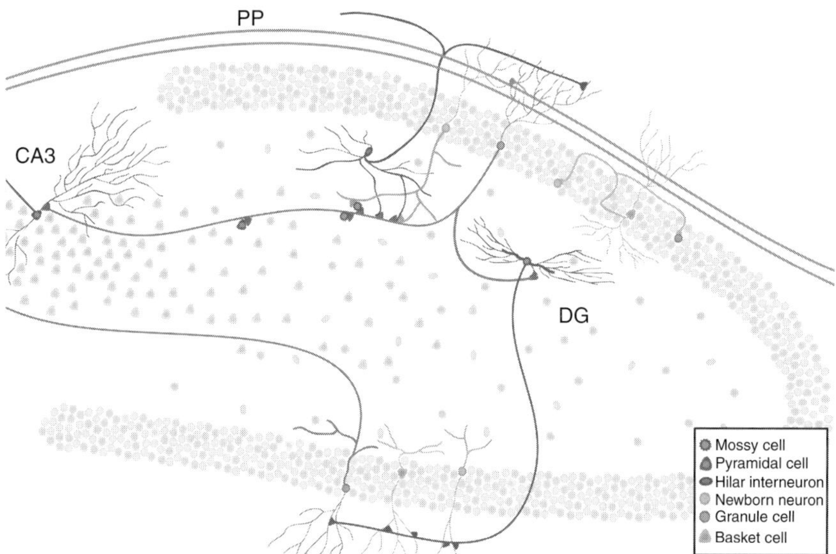

Figure 5.2 The cells and connections in the dentate gyrus.

2. Development of the Hippocampal Circuit

Stem cells, and progenitor cells of both pyramidal neurons and DGCs originate from the ventricular germinal layers (Altman and Bayer, 1990b). The hippocampal neuroepithelium consists of three morphologically different components: the ammonic neuroepithelium, primary dentate neuroepithelium, and fimbrial glioepithelium. These components generate the pyramidal neurons of the hippocampus, the granular neurons of the dentate gyrus, and the glial cells of the fimbria, respectively (Bayer, 1980a). Pyramidal neurons of the hippocampus are generated between E10 and E18 in the mouse and between E16 and E21 in the rat (Bayer, 1980b). The early-generated pyramidal cells are destined to settle in the late-forming stratum pyramidale of CA3 region, whereas the late-generated pyramidal cells settle in the early-forming CA1 region. The reverse order in the settling and generation of pyramidal cells might occur because CA3 pyramidal cells have to wait for the formation of the DGC layer, which, in terms of both neurogenesis and morphogenesis, is the latest-developing component of the hippocampus. The generation of granular neurons in the dentate gyrus starts from E16, and continues through adulthood. Although a few DGCs are generated quite early, the dentate gyrus is not recognizable as a morphological structure with even partially formed external limbs along the hippocampal fissure until E21. And, CA3 pyramidal cells are generated several days earlier, but they have to pause in the intermediate zone until some of the DGCs, the source of their afferents, settle in the granular layer (Altman and Bayer, 1990b).

The putative primary dentate neuroepithelium is distinguished from the hippocampal neuroepithelium by its shape and cellular dynamics (Bayer, 1980a). It is located around the dentate notch; its cells protrude in the direction of the pial surface (Altman and Bayer, 1990a). By E18, the dentate neuroepithelium is surrounded by a secondary germinal matrix. The primary dentate neuroepithelium around the dentate notch is gradually reduced in size and the expanding fimbria comes to occupy the vacated area. The continuity between the primary dentate neuroepithelium and the secondary dentate matrix is maintained until E20; thereafter, the dentate notch is no longer related to the dentate gyrus but to the fimbria. By day E21, there is a massive cell migration from the secondary matrix to the crest of the dentate gyrus, and densely packed cells are found except in the external limb of the DGC layer. By day E22, the external limb becomes longer and more distinct, and the internal limb of the DGC layer starts to form as an extension of the crest. At the same time, two components of the dentate migration become distinguished, although fewer cells are found in the secondary matrix. The first dentate migration follows a subpial route and adds cells to the expanding internal limb of DGC layer. The secondary migration is situated beneath the pyramidal cells of the hippocampus and above the internal limb. The first dentate migration disappears at postnatal (P) day 10; however, the secondary dentate migration is present until P30. Between days P20 and P30, proliferating cells become confined to the subgranular zone (SGZ) at the base of the granular layer, which is the source of DGCs produced in the adult hippocampus.

The terminals of entorhinal fibers arrive at their termination of the outer two-third of the dentate molecular layer and stratum lacunosum-moleculare of hippocampus at day E17 (Super and Soriano, 1994). The early-generated Cajal–Retzius cells locate in the marginal zone and project to the entorhinal cortex as early as E17 (Ceranik *et al.*, 1999). It has been suggested that the axons of Cajal–Retzius cells that project from the hippocampus to the entorhinal cortex provide a template or a guiding scaffold for the outgrowing axons of the entorhinal cortex (Frotscher, 1998). Cajal–Retzius cells might be the targets of entorhinal projection neurons in the early stage, since the distal dendrites of granular cells and pyramidal neurons have not reached the outer layer of the dentate molecular layer and the stratum lacunosum-moleculare of the hippocampus by that time (Del Rio *et al.*, 1997; Frotscher *et al.*, 2001). During postnatal development, the connections of the entorhinal cortex are reorganized and the final connections are formed after the degeneration of Cajal–Retzius cells.

Commissural projections originate from CA3 pyramidal neurons and hilar mossy cells and arrive at the contralateral hippocampus and dentate gyrus through the hippocampal commissure. The commissural axons develop later than the entorhinal ones (Bayer and Altman, 1987). When the commissural axons arrive at the contralateral hippocampus, their target neurons, hippocampal pyramidal cells, and DGCs have already been generated and the pioneer target neurons are not required (Deller *et al.*, 1999; Zhao *et al.*, 2003a), in striking contrast to the entorhinal axon projection, as described above (Forster *et al.*, 2006).

3. Producing New Neurons from Adult Neural Stem Cells in the Hippocampus

In the adult hippocampus, new neurons are constantly generated from a resident population of stem cells located in the border of the hilus and granule cell layer of the dentate gyrus. These neural stem cells (NSCs) are defined by their ability to undergo multipotent differentiation into major cell lineages, including neurons and astrocytes, as well as by their capacity to produce identical cells (self-renew) (Gage, 2000). *In vitro*, NSCs or stem-like cells have been isolated from the dentate gyrus (Song *et al.*, 2002). A recent *in vivo* fate analysis study of Sry-related HMG-box transcription factor *Sox2* transgenic mice showed that a single *Sox2*-positive cell can generate neurons, astrocytes, and identical $Sox2^{+}$ cells at single-cell level, providing the first *in vivo* evidence of the multipotent and self-renewal properties of NSCs in the adult hippocampus (Suh *et al.*, 2007).

Generating new neurons is a complex, multiple-step process that originates from neural progenitor cells (Kempermann *et al.*, 2004; Zhao *et al.*, 2008). Corresponding to the multiple-step process, several types of intermediate precursor cells can be identified according to their specific morphologies, electrophysiological properties, and expression of unique molecular markers in the dentate gyrus in adult hippocampus (Figs 5.3 and 5.4) (Filippov *et al.*, 2003; Seri *et al.*, 2004). A radial glia-like precursor cell is infrequently labeled by BrdU, and has been called a type 1 cell. The cell body of these cells is located in the SGZ, the soma is triangular, and its long process reaches through the granule cell layer and into the inner molecular layer. These cells express an intermediate filament protein Nestin, an astrocytic marker glial fibrillary acidic protein (GFAP), a radial glia marker BLBP, and *Sox2*. In the subventricular zone (SVZ), another region maintaining neurogenesis property in the adult brain, GFAP-positive cells also express astrocyte-specific glutamate transporter (GLAST) (Liu *et al.*, 2006; Sundholm-Peters *et al.*, 2004). While in the SGZ, GLAST is not a widely used marker and more experimental results are needed (Namba *et al.*, 2005). These radial cells exhibit the electrophysiological character of astrocytes: expressing passive, noninactivating currents with a linear current–voltage relationship, and having a reversal potential close to the K^+ equilibrium potential (Filippov *et al.*, 2003; Fukuda *et al.*, 2003; Steiner *et al.*, 2006). Type 2 cells have a distinct morphology: smaller cell body, irregularly shaped dense nucleus and short, horizontal processes without long radial processes. Cells with these morphological characteristics are more

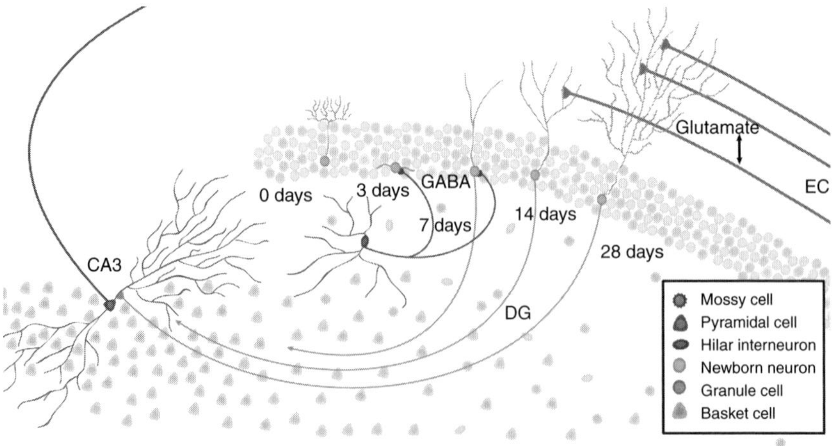

Figure 5.3 The time course of neurogenesis in the adult dentate gyrus.

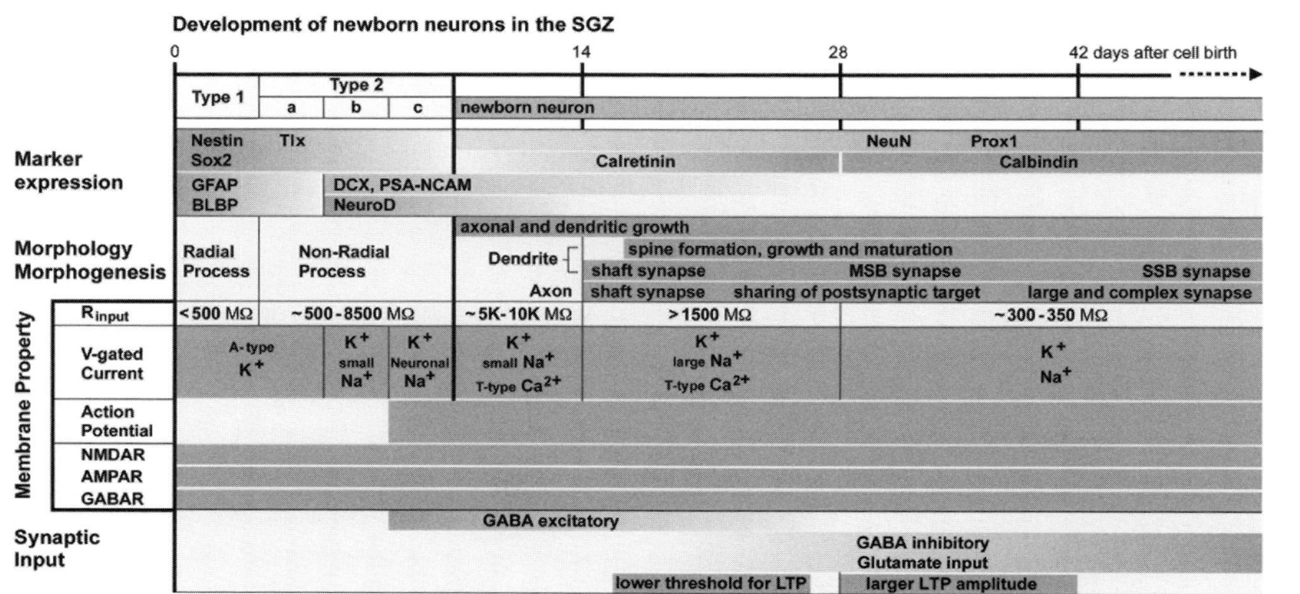

Figure 5.4 Development of newborn neurons in the SGZ.

heterogeneous (Fukuda *et al.*, 2003; Steiner *et al.*, 2006; Wang *et al.*, 2000). Type 2a cells express delayed-rectifier K^+ currents; they resemble type 1 cells except that they lack long radial processes (Steiner *et al.*, 2006). Type 2b cells are BLBP- and GFAP-negative but doublecortin (DCX) or polysialylated embryonic form of neural cell adhesion molecule (PSA-NCAM) positive cells (Fukuda *et al.*, 2003; Steiner *et al.*, 2006). These cells have high-input resistance and voltage-dependent Na^+ currents. The third type 2 cell is characterized by the typical inward neuronal Na^+ currents. Action potentials can be generated when depolarizing current injected into these cells, which is an early sign of neuronal differentiation (Wang *et al.*, 2005). Nonradial cells mark a progression from stem-like progenitor cells to precursors of neuronal lineage in the adult hippocampus. Primary cilia, which mediate Sonic hedgehog (Shh) signaling, are found on both radial and nonradial precusor cells in the DG. Ablation of the primary cilia disrupted the expansion and development of postnatal hippocampal progenitors through the Shh signal (Breunig *et al.*, 2008; Han *et al.*, 2008).

The proliferation and survival of NSCs is a complex process and many steps of which are affected by the electrical activity of hippocampus. It has been demonstrated that NSCs in the hippocampus can directly sense the excitatory stimuli through L-type Ca^{2+}-channels and NMDA receptors to increase the neurogenesis both *in vitro* and *in vivo*. The excitation inhibits the expression of glia fate gene *Hes1* and *Id2* while increasing *NeuroD* expression to promote neuronal differentiation (Deisseroth *et al.*, 2004).

The first functional synaptic innervations from the hippocampal circuitry on progenitor cells are GABAergic. Both types of precursor cells express functional $GABA_A$ and glutamate receptors (Tozuka *et al.*, 2005; Wang *et al.*, 2005). However, only type 2 cells receive synaptic inputs, and these inputs are GABAergic, not glutamatergic (Tozuka *et al.*, 2005; Wang *et al.*, 2005). The application of GABA produces a membrane depolarization and triggers an increase in $[Ca^{2+}]_i$ in type 2 cells. In the short-term (20-h) hippocampal slice culture, the calcium influx induced by $GABA_A$ receptor activation increases NeuroD expression. The excitation of type 2 cells by GABAergic inputs promotes their neuronal differentiation and consequently enhances adult neurogenesis in the hippocampus.

4. Maturation and Integration of Newly Generated Cells in the Hippocampal Circuitry

NSCs undergo discrete developmental stages—proliferation, differentiation, survival, migration, maturation, and integration—to finally generate functional new neurons. The entire process requires about 4 weeks and progresses through a similar cascade of events as during hippocampal development (Fig. 5.3). Retrovirus-mediated, enhanced green fluorescent protein expression allows accurate morphological and electrophysiological analyses of adult-born neurons (van Praag *et al.*, 2002; Zhao *et al.*, 2006).

Most of GFP^+ cells are located at the border of the granule cell layer at 3 days postretrovirus injection (dpi). A small portion of the cells migrates to the middle third of the DGC layer. Some of the processes are parallel to the granule cell layer and a few extend to the DGC layer. At 7 dpi, most of the cells have processes spanning the DGC layer. Some GFP^+ cells cluster together at 3 and 7 dpi. At this stage, neurons lack synaptic inputs. All properties are typical of immature neurons in the developing and adult brain.

At 14 dpi, GFP^+ cells are already migrating into the granule cell layer. The apical dendrites reach the middle molecular layer and their dendritic arborizations become more elaborate, but no spines are observed on the dendrite. In most cases, basal dendrites, which will retract at a late stage of maturation, are also observed and project to the hilus at this stage. Parallel with morphological maturation, excitability increases and GABAergic afferents are observed. Neurons with GABAergic but not glutamatergic input are first observed at 8 dpi. The kinetics of the $GABA_A$-mediated postsynaptic current is the slow dendritic response rather than the fast perisomatic current. During maturation, E_{GABA} gradually decreases in DGCs, indicating that GABA initially depolarizes the newborn cells in the adult brain, which is similar to that in the developing brain. Two weeks after birth, neurons start to resemble the typical morphological and physiological properties of DGCs, although they still have the characteristics of immature neurons.

GFP^+ cells at 28 dpi show mature features. These neurons display round soma and a conspicuous axonal projection extending toward the hilus to the CA3 region of hippocampus; their spiny dendrites reach the outer molecular layer. Repetitive spikes with large amplitude and high-frequency adaptation are elicited by depolarizing currents, which are typical characteristics of mature DGCs. Glutamatergic synaptic inputs are detected at this stage and the early glutamatergic current recorded from newborn DGCs displays a lower threshold for long-term potentiation (LTP) induction. Fast kinetics GABAergic postsynaptic currents are observed only in neurons bearing functional glutamatergic inputs. In brief, the development of afferent synaptic connections to newborn cells in the adult hippocampus takes about 4 weeks and follows a sequence similar to hippocampal development: no input–GABA excited input–glutamate excited input–GABA inhibitory input (Fig. 5.4).

Fine morphology studies using electron tomography and serial section electron microscopy (EM) have shown that new neurons at 30 dpi are similar to mature neurons and receive a variety of inputs: axosomatic, axodendritic, and axospinous (Fig. 5.2). Some of the axospinous afferents originate from entorhinal cortex, as revealed by tracing experiments (Toni *et al.*, 2007). More abundant dendritic filopodia are found during the early stage of maturation. The size and density of dendritic protrusions increase with the time of development. In three-dimensional analysis, the tips of all filopodia preferentially associate with axon terminals, which already form synapses with dendritic spines from other neurons. Spines of new neurons contact with multiple-synapse boutons (MSB) at the beginning; as neurons mature, spines form synapses only with single boutons devoid of the other synaptic partners and become single-synapse boutons (SSB) (Figs 5.2 and 5.4). Although the first glutamatergic synapse appears during the third week after division, full maturity of the excitatory input takes 60 days.

The morphological and electrical development of newly generated cells in the adult hippocampus follows a similar transition process. However, the maturation processes of adult-generated granules are slower than those of neonate-generated DGCs (Overstreet-Wadiche *et al.*, 2006; Zhao *et al.*, 2006). GFP cells identified in proopiomelanocortin (POMC)-EGFP transgenic mice display the immature morphological and physiological properties of DGCs in the dentate gyrus. Using *BrdU* labeling to compare when neonatal and adult DGCs achieve the GFP expression stage in POMC-EGFP mice, it was found that, in the adult, colabeling between *BrdU* and GFP peaked at 12 days after *BrdU* injection and declined gradually over 2 weeks; however, in the neonate, colabeling peaked at 8 days and declined more rapidly (Overstreet-Wadiche *et al.*, 2006). If we compare colabeling between *BrdU* and the mature neuronal marker, NeuN, at 12 days after *BrdU* injection, the percentage of colabeling in the adult dentate is significantly lower than in the neonatal dentate. Moreover, the initiation of spine growth in the newly generated granule cells in the adult labeled with retrovirus is delayed by around 4 days compared to the neonate. The dendritic length of adult-generated neurons is markedly less than that of the neonate (Zhao *et al.*, 2006). Although the maturation of granule cells generated in the adult brain is delayed, DGCs generated in the neonate and adult constitute a functionally homogeneous population that receives remarkably similar afferent connectivity (Laplagne *et al.*, 2006, 2007) in the adult brain.

Axons of granule cells in the dentate gyrus, called mossy fibers, project into the CA3 regions of hippocampus. The process can be detected as early as 10 days after cell birth in the adult brain (Hastings and Gould, 1999; Zhao *et al.*, 2006). Retrovirus labeling studies show that the axonal and dendritic outgrowths are concomitant and axons reach the CA3 region before the first spines are formed in the newly generated neurons. Similar to dendritic growth, axonal growth is delayed in the adult brain. Recent studies have demonstrated functional synapse formation by newly generated DGCs in the adult (Toni *et al.*, 2008). GFP-positive axons are found forming synapses on the target cells in the CA3 area and hilus. The average size of the mossy fiber boutons in CA3 is significantly larger than that in the hilus after 17-dpi retrovirus. The larger mossy fiber terminals form synapses with thorny excrescences and spiny dendrites in the CA3 area and with thorny excrescences in the hilus. En passant synapses with thin, aspiny dendritic shafts are also found, suggesting that GFP-positive axons form synapses with interneurons. At 17 dpi, GFP-positive mossy fiber boutons form synapses with the dendritic shaft of pyramidal cells. At 28 dpi, GFP-positive mossy fiber boutons form synapses on the thorny excrescences that already have synapses with GFP-negative boutons. At 75 dpi, GFP-positive boutons are entirely associated with one individual thorny excrescence. During maturation, the presynaptic vesicle significantly increases from 60 at 17 dpi to 179 at 75 dpi. By using retroviral expression channelrhodopsin-2 (ChR2) (Zhang *et al.*, 2006) in newborn DGCs, functional synaptic responses induced by blue light are detected in the target cells in the CA3 and hilus. Taken together, EM and physiological studies show that axons of newborn DGCs can form functional synapses with hilar interneurons, mossy cells, and CA3 pyramidal cells (Fig. 5.2).

5. Functional Implications of Neurogenesis in the Adult Hippocampus

Hippocampal cells born during adulthood participate in the preexisting neural circuit; they receive functional excitatory and inhibitory synaptic inputs from afferent fibers similar to those of neurons generated during early development (Laplagne *et al.*, 2006; van Praag *et al.*, 2002). Given the critical role of the hippocampus in learning and memory, whether and how adult neurogenesis participates in hippocampal functions have been central questions. In this review, we summarize and discuss the evidence for reciprocal relations between learning and neurogenesis in the adult hippocampus.

5.1. From cell to network

At the cellular level, the progression of the establishment of synaptic inputs to adult-born DGCs in mice has been fully characterized using electrophysiology. The newborn neurons lack afferent synaptic contacts but receive tonic GABA activation. They begin to receive GABAergic synaptic inputs of dendritic origin around 1 week after birth and then glutamatergic inputs by 2 weeks. The synaptic integration is finally completed by the fourth week with the formation of perisomatic GABAergic contacts (Ambrogini *et al.*, 2004a; Esposito *et al.*, 2005; Ge *et al.*, 2006; Karten *et al.*, 2006; Overstreet Wadiche *et al.*, 2005; Wang *et al.*, 2005). This maturation process in the adult brain has striking similarities to that occurring in neonates. However, analyses of POMC-EGFP and retrovirus-labeled mice have shown that DGCs of identical postmitotic age have greater dendritic length, higher spine density, and more mature, excitable membrane properties in neonates than in adults (Overstreet-Wadiche *et al.*, 2006; Zhao *et al.*, 2006), indicating that the development of neurons born in adults is significantly delayed in comparison with those born during early postnatal development.

The young DGCs differ substantially from their older, neighboring counterparts in terms of both active and passive membrane properties (Schmidt-Hieber *et al.*, 2004; Wang *et al.*, 2000). For instance, they have high-input resistance and activation of low-threshold T-type Ca^{2+}-channels under physiological conditions, which lead to enhanced excitability and generation of action potentials in response to even weak excitatory inputs. Furthermore, associative LTP can be induced more easily in young neurons than in mature ones under identical conditions (Schmidt-Hieber *et al.*, 2004). This reduced threshold for LTP induction might result from the lack of fast inhibitory GABA-mediated synaptic inputs to newborn neurons (Snyder *et al.*, 2005; Wang *et al.*, 2000) and/or from the low Ca^{2+}-buffering capacity caused by low expression of calbindin in them (Muller *et al.*, 2005). However, adult-born neurons exhibit enhanced synaptic plasticity with both increased LTP amplitude and decreased LTP induction threshold only within a fairly narrow time window, between 1 and 1.5 months of cell age, depending on activation of NR2B-containing NMDA receptors (Ge *et al.*, 2007). The maintenance of late phase LTP and the consolidation of long-lasting memory require expression of certain immediate early genes, such as Zif268 and Arc/Arg3.1 (Bozon *et al.*, 2002; Chawla *et al.*, 2005; Guzowski *et al.*, 2000). Consistent with the electrophysiology studies described above, LTP is found to induce Zif268 expression in a substantial fraction of 2-week-old neurons in adult rats (Bruel-Jungerman *et al.*, 2006). In addition, preferential Arc expression in newly generated DGCs rather than their old neighbors can be induced by activity associated with spatial exploration, memory formation, or memory retrieval (Kee *et al.*, 2007; Ramirez-Amaya *et al.*, 2006). Taken together, these studies suggest that neurogenesis in the adult hippocampus continuously assembles ensembles of new neurons transiently exhibiting enhanced excitability and synaptic plasticity. These unique properties may enable newborn neurons to function differently from their older neighbors, at least within a certain time window. However, there is not yet definitive evidence that adult-born cells constitute a specialized population to fulfill specific functions.

In contrast to the comprehensive characterizations of afferent connectivity, the outputs of newborn neurons have rarely been examined by electrophysiological methods because of the technical difficulty in finding pairs of newborn neurons and their postsynaptic targets. A recent study overcame this problem by selectively introducing light-activated cation channel ChR2 into newly generated DGCs and simultaneously activating a large number of them by blue light, which increases the possibility of finding their postsynaptic targets through whole cell recording from randomly selected neurons in the hilus and CA3 region (Toni *et al.*, 2008). The results suggest that DGCs generated in the adult dentate gyrus send functional synaptic projections to CA3 pyramidal cells as well as to hilar interneurons and release glutamate as the main neurotransmitter. Although early studies have suggested that a small fraction of neural progenitors in the adult rat dentate gyrus may also become GABAergic basket cells (BCs) and form inhibitory synapses with the principal DGCs (Liu *et al.*, 2003), other studies have not observed incorporation of either BrdU or [^{3}H]-thymidine into BCs (Seri *et al.*, 2004). Therefore, whether neurotransmitters other than glutamate can be released by adult-born neurons needs to be further confirmed.

At the network level, newborn neurons might bring special properties to the local microcircuits that they join by concerted activities of neurons within the circuits. They may integrate into the pre-existing network with or without changing the characteristics of their mature afferent and efferent neurons, or they may simply replace their old dying neighbors. In either case, they may give rise to unique activity patterns of the neuronal ensembles depending on their electrophysiological stage of maturation (Lledo *et al.*, 2006). New DGCs generated in the pathological context of seizures show certain properties that are different from those produced under normal physiological conditions, including reduced excitability as well as increased inhibitory and reduced excitatory synaptic drive, suggesting that these new cells may help to keep a homeostasis in overall network activity (Jakubs *et al.*, 2006). Nevertheless, there has been no direct evidence demonstrating the contribution of newborn neurons in the context of the activity of the entire network. Since neuronal synchronization and oscillation in different frequency domains induce specific forms of cellular plasticity during subsequent stages of memory formation (Axmacher *et al.*, 2006), it would be interesting to find out whether and how these two features can be altered by integration of new neurons and vice versa.

5.2. Regulation of neurogenesis by activity

The basal rate of neurogenesis or the proliferation of granule cells in the adult hippocampus is believed to be genetically determined (Cameron and McKay, 2001). However, the process of neurogenesis is highly regulated by various physiological stimuli (Table 5.1). It was originally discovered that, in adult black-capped chickadees, seasonal differences in the recruitment of newly generated hippocampal neurons positively correlated with seasonal changes in spatial learning-related behaviors (Barnea and Nottebohm, 1994). Consistently, birds that engage in spatial learning behaviors, such as parasitic brooding and long-distance migration, seem to have a larger

Table 5.1 Regulation of hippocampal neurogenesis by experience

Experience	Effects on neurogenesis	References
Enriched environment	Survival +	Kempermann *et al.* (1997a) and Nilsson *et al.* (1999)
Running	NPC proliferation +	van Praag *et al.* (1999)
Stress	NPC proliferation −	Gould *et al.* (1997, 1998) and Tanapat *et al.* (2001)
Water maze	Survival +	Ambrogini *et al.* (2000), Gould *et al.* (1999), and Hairston *et al.* (2005)
Trace eyeblink	Survival +	Gould *et al.* (1999) and Leuner *et al.* (2004)
Water maze (late phase)	NPC proliferation +; survival ± depending on born time	Dobrossy *et al.* (2003)

Abbreviations: +, increase; −, decrease.

hippocampus (Healy *et al.*, 1996; Sherry *et al.*, 1993). In the case of adult mammals, retrovirus-mediated single-cell gene knockout of NMDA receptors in mice revealed that the survival of individual new neurons is regulated in an activity-dependent manner (Tashiro *et al.*, 2006). At the system level, several studies have shown that enriched environment enhances the survival of newly generated cells (Kempermann *et al.*, 1997b; Nilsson *et al.*, 1999), whereas voluntary exercise increases the proliferation of neural precursor cells (NPCs) in the dentate gyrus (van Praag *et al.*, 1999). In contrast, stressful experience decreases the number of new DGCs, generally by down-regulating cell proliferation rather than by altering cell survival. For instance, predator odor exposure in adult rats and social stress in tree shrews and marmosets inhibit the proliferation of NPCs in the dentate gyrus (Gould *et al.*, 1997; Tanapat *et al.*, 2001). Specific learning paradigms, such as place learning in a Morris water maze and trace eyeblink conditioning, increase the number of new neurons born 1 week prior to training (Ambrogini *et al.*, 2000; Gould *et al.*, 1997, 1998; Hairston *et al.*, 2005) but do not alter the number of those cells generated during training (van Praag *et al.*, 1999), suggesting that learning-related neuronal activities facilitate the integration of new neurons into existing circuitry and their survival during the time when their axons and dendritic trees are still under development (Zhao *et al.*, 2006). The observed increase in neurogenesis is specific to hippocampus-dependent learning, since delayed eyeblink conditioning and cue learning in the water maze, which do not require the integrity of hippocampus, do not modify neurogenesis (Gould *et al.*, 1998). Besides, the newborn cells rescued from death by learning survive for months after memory formation, and the performance of individual animals positively correlates with cell survival (Leuner *et al.*, 2004), indicating that learning rather than training is sufficient for enhancing the lifespan of newly generated neurons.

Recent evidence suggests that the late phase of spatial learning in a Morris water maze may also induce the proliferation of NPCs in the dentate gyrus (Dobrossy *et al.*, 2003). Consistent with behavioral studies, associative LTP, which is regarded as the mechanism for memory formation and retention, has been shown to stimulate the proliferation of progenitor cells in the dentate gyrus and to promote survival of 1- to 2-week-old DGCs (Bruel-Jungerman *et al.*, 2006). On the other hand, spatial learning may decrease the number of newborn cells by inducing death of immature neurons, depending on the time elapsing from mitosis to learning, which might constitute a trimming mechanism to suppress neurons that have not established learning-related synaptic connections (Ambrogini *et al.*, 2004b; Dobrossy *et al.*, 2003; Epp *et al.*, 2007). In summary, learning modifies neurogenesis through enhancement of a cascade of interrelated events, including proliferation of NPCs, survival of relatively mature neurons, and apoptosis of relatively immature cells, suggesting that neuronal activities sculpt neural networks through a highly selective addition and suppression of distinct populations of newborn neurons.

5.3. Impact of neurogenesis on learning and memory

It is generally accepted that the hippocampus serves a crucial role in episodic memory (Neves *et al.*, 2008). Although there has been no direct evidence demonstrating whether and how adult-generated hippocampal cells help to achieve this function, numerous studies have suggested a strong link between the rate of neurogenesis and learning capabilities.

The basal level of neurogenesis in rodents varies with genetic background (Kempermann *et al.*, 1997a, 1998b). Acquisition of spatial memory in the water maze positively correlates with the neurogenesis level of the strain (Kempermann and Gage, 2002). This finding is consistent with a quantitative relationship between the performance of individual aged animals on a hippocampus-dependent task and the number of new neurons generated by them (Bizon and Gallagher, 2003; Drapeau *et al.*, 2003). Similarly, enhanced neurogenesis induced by enriched environment or physical activity is associated with improved performance on spatial navigation learning tasks (Kempermann *et al.*, 1997b, 1998a; van Praag *et al.*, 1999), even in middle-aged or senescent mice (Kempermann *et al.*, 1998b; van Praag *et al.*, 2005). In contrast, aging (Drapeau and Nora Abrous, 2008), stress (Lemaire *et al.*, 2000; Mirescu and Gould, 2006), and cholinergic lesions (Mohapel *et al.*, 2005) decrease both neurogenesis and performance on hippocampus-dependent learning tasks. Nevertheless, the possibility that neurogenesis and learning are independently regulated by the experimental manipulations cannot be ruled out, as indicated by the observation by one group, that environmental enrichment alters spatial learning and anxiety-like behavior in mice when hippocampal neurogenesis is blocked by X-irradiation (Meshi *et al.*, 2006). Therefore, experiments designed to examine the effects on behaviors of specifically manipulating adult neurogenesis are required to directly test whether new hippocampal neurons play a role in learning. The rationale is that, if neurons born in adulthood are necessary for a certain hippocampus-dependent task, reduced neurogenesis should be associated with impaired learning, whereas enhanced neurogenesis should improve learning. As summarized in Table 5.2, this hypothesis is supported by studies using various approaches, although the results are still debatable depending on the experimental situation.

Two major methods to reduce the number of adult-generated neurons are irradiation and administration of the antimitotic agent, methylazoxymethanol acetate (MAM). Low doses of cranial ionizing radiation abolish young neuron-mediated LTP in the dentate gyrus (Snyder *et al.*, 2001). This particular form of LTP is independent of GABAergic inhibition; therefore, it might be important especially during active behavioral exploration, when inhibition in the hippocampus is at a high level (Moser, 1996). Irradiation also results in significant impairment of the learning tasks that are known to require the hippocampus, such as place recognition in a T-maze, spatial learning in the Barnes maze, contextual fear conditioning, and nonmatching-to-sample (NMTS) task with long intervals between sample and test trials, whereas spatial learning in the water maze and hippocampus-independent learning (e.g., object-recognition task) are unaffected (Madsen *et al.*, 2003; Meshi *et al.*, 2006; Raber *et al.*, 2004; Saxe *et al.*, 2006; Snyder *et al.*, 2005; Winocur *et al.*, 2006). Besides, X-irradiation blocks the behavioral effects of antidepressant treatment, which generally increases adult hippocampal neurogenesis (Santarelli *et al.*, 2003). Likewise, the mice treated with MAM perform more poorly than control animals in trace eyeblink conditioning and trace fear conditioning (Shors *et al.*, 2001, 2002). MAM treatment also prevents long-term memory improvement induced by environmental enrichment (Bruel-Jungerman *et al.*, 2005). Despite these compelling correlative pieces of evidence, the role of adult-born neurons in learning remains inconclusive due to the nonspecific nature of both manipulations. Irradiation can cause persistent inflammatory response and MAM can reduce cell proliferation systemically (Dupret *et al.*, 2005; Monje *et al.*, 2002). Thus, their influence is not exclusively limited to the hippocampus and might cause side effects unrelated to the reduction in neurogenesis. Genetic modifications, including gene knockout and transgenic techniques, have recently been used to manipulate adult neurogenesis, too. For instance, development of inducible GFAP-thymidine kinase (TK) transgenic mice provides a new method for ablating the hippocampal precursor population (Garcia *et al.*, 2004), and an impairment in contextual fear conditioning has been observed in such a mouse line

Table 5.2 Influence of hippocampal neurogenesis on learning

Regulators	Neurogenesis	Effects on behavior	References
Enriched environment	+	Water maze ↑	Kempermann *et al.* (1997a, 1998a)
Running	+	Water maze ↑	van Praag *et al.* (1999)
Aging	−	Water maze ↓	Bizon and Gallagher (2003) and Drapeau *et al.* (2003)
Stress	−	Water maze ↓	Lemaire *et al.* (2000)
Irradiation	−	Place recognition ↓	Madsen *et al.* (2003)
		Barnes maze ↓	Raber *et al.* (2004)
		Water maze ↓	Snyder *et al.* (2005)
		Contextual fear conditioning ↓	Saxe *et al.* (2006)
		NMTS task ↓	Winocur *et al.* (2006)
MAM	−	Trace eyeblink conditioning ↓	Shors *et al.* (2001)
		Trace fear conditioning ↓	Shors *et al.* (2002)
		Novel object-recognition task ↓	Bruel-Jungerman *et al.* (2005)
Ablation of GFAP progenitors	−	Contextual fear conditioning ↓	Saxe *et al.* (2006)
Deletion of MBD1	−	Water maze ↓	Zhao *et al.* (2003a,b)
VEGF overexpression	+	Water maze ↑	Cao *et al.* (2004)
Removal of TLX	−	Water maze ↓	Zhang *et al.* (2008)
Bax overexpression	−	Water maze ↓	Dupret *et al.* (2008)

Abbreviations: +, increase; −, decrease; ↑, improved; ↓, impaired.

(Saxe *et al.*, 2006). In another example, mice lacking methyl-CpG-binding domain protein 1 (MBD1) have reduced hippocampal neurogenesis associated with impaired spatial learning and a significant reduction in LTP within the dentate gyrus (Zhao *et al.*, 2003b), whereas hippocampal gene transfer of vascular endothelial growth factor (VEGF) in adult rats leads to enhanced neurogenesis associated with improved cognitive function in the water maze and passive avoidance tasks (Cao *et al.*, 2004). Furthermore, the presenilin-1 (PS1) conditional knockout mice exhibited deficiency in enrichment-induced neurogenesis, which is accompanied by prolonged long-term memory retention in the contextual fear-conditioning test, suggesting that new neurons might contribute to periodic clearance of outdated hippocampal memory traces after cortical memory consolidation (Feng *et al.*, 2001). However, the genetic modifications described above still lack specificity in controlling neurogenesis, for example, tissues other than hippocampus or cells other than newborn neurons might be affected, autonomous compensatory mechanisms cannot be excluded, etc. Recently, Zhang *et al.* (2008) generated transgenic mice in which orphan nuclear receptor TLX could be specifically removed in adults through inducible recombination to reduce adult NSC proliferation. The suppression of adult neurogenesis was found to result in a marked decrement in spatial learning but not in contextual fear conditioning. Dupret *et al.* (2008) developed another conditional transgenic strategy that allows specific ablation of newborn hippocampal neurons in adult animals. This animal model is based on selective overexpression of the proapoptotic protein Bax in Nestin-expressing precursors in an inducible manner. Removal of newborn cells in the adult dentate gyrus causes an impairment of spatial relational memory, whereas simpler forms of spatial knowledge and memory that are independent of the hippocampus are spared. Both studies provide direct support that there is a causal relationship between adult neurogenesis and learning.

5.4. Computational modeling

Due to the intrinsic complexity of hippocampal neural circuits, artificial neural networks consisting of several relatively simple and interconnected computational units serve as models in which adult neurogenesis can be specifically controlled. Computational studies on these simplified systems have theorized possible functions for continuously adding new neurons into the hippocampus. By using a three-layer neural network and assuming neuronal turnover or coordinated apoptotic and neurogenic events in the middle layer, Chambers *et al.* (2004) and Deisseroth *et al.* (2004) have found that incorporation of new neurons facilitates both the clearance of old memories and the storage of new memories. Simulations of the complete hippocampal circuit (Becker, 2005) suggest that neural turnover in the dentate gyrus is used to change the encoding over time and thereby contributes to creation of distinct memory traces for highly similar patterns. This result also supports the prediction that memory capacity increases with the number of DGCs. However, simply adding new excitatory neurons to a network may lead to epileptiform processes and impaired memory recall, which are proposed to be avoided by circuit-level homeostatic mechanisms to compensate for increased hippocampal activity (Meltzer *et al.*, 2005). Considering the finding that new neurons are more plastic than old neighboring cells in terms of encoding new information, Wiskott *et al.* (2006) proposed that new neurons help to avoid the problem of catastrophic interference when adapting to new environments. The special physiological properties of immature neurons are also hypothesized to alter the sparse coding function of the dentate gyrus, which could possibly be the source of temporal associations in memory (Aimone *et al.*, 2006). Moreover, a recent study suggested that immature new neurons introduce a degree of similarity to memories learned at the same time whereas fully mature new neurons improve encoding of future memories (Aimone *et al.*, 2009).

6. Concluding Remarks

Based on the studies conducted over the past decade, it has become more and more clear that neurogenesis in the adult hippocampus, a special form of neural circuit development persisting in the mature brain, is tightly linked to cognitive and emotional behaviors. It has been well recognized that adult-born DGCs transiently exhibit unique properties during maturation, such as enhanced excitability and plasticity. However, the real physiological significance of these properties and the role of fully mature new neurons in memory are much less clear. To address these issues, future research aimed at identifying specific contributions of newborn neurons at different maturation stages will be critical. In other words, it is of vital importance to find out the exact role of immature and mature new neurons in the formation, consolidation, and retrieval of long-term memory. In this regard, refined strategies are needed to enable control of adult-born cells in the dentate gyrus exclusively and at specific time points. Furthermore, it is necessary to develop behavioral tests that have better sensitivity and consistency. It should also be noted that a better understanding of the continuous development of hippocampal neural circuits in adulthood requires a better understanding of the function of the dentate gyrus and the hippocampus itself.

REFERENCES

Aimone, J. B., Wiles, J., and Gage, F. H. (2006). Potential role for adult neurogenesis in the encoding of time in new memories. *Nat. Neurosci.* **9,** 723–727.

Aimone, J. B., Wiles, J., and Gage, F. H. (2009). Computational influence of adult neurogenesis on memory encoding. *Neuron.* **61,** 187–202.

Altman, J., and Bayer, S. A. (1990a). Migration and distribution of two populations of hippocampal granule cell precursors during the perinatal and postnatal periods. *J. Comp. Neurol.* **301,** 365–381.

Altman, J., and Bayer, S. A. (1990b). Prolonged sojourn of developing pyramidal cells in the intermediate zone of the hippocampus and their settling in the stratum pyramidale. *J. Comp. Neurol.* **301,** 343–364.

Ambrogini, P., Cuppini, R., Cuppini, C., Ciaroni, S., Cecchini, T., Ferri, P., Sartini, S., and Del Grande, P. (2000). Spatial learning affects immature granule cell survival in adult rat dentate gyrus. *Neurosci. Lett.* **286,** 21–24.

Ambrogini, P., Lattanzi, D., Ciuffoli, S., Agostini, D., Bertini, L., Stocchi, V., Santi, S., and Cuppini, R. (2004a). Morpho-functional characterization of neuronal cells at different stages of maturation in granule cell layer of adult rat dentate gyrus. *Brain Res.* **1017,** 21–31.

Ambrogini, P., Orsini, L., Mancini, C., Ferri, P., Ciaroni, S., and Cuppini, R. (2004b). Learning may reduce neurogenesis in adult rat dentate gyrus. *Neurosci. Lett.* **359,** 13–16.

Andersen, P. (1975). Organization of hippocampal neurons and their interconnections. *In* "The Hippocampus" (R. L. Isaacson and K. H. Pribram, eds.), vol. 1, pp. 155–174. Plenum Press, New York.

Andersen, P., Blackstad, T. W., and Lomo, T. (1966a). Location and identification of excitatory synapses on hippocampal pyramidal cells. *Exp. Brain Res.* **1,** 236–248.

Andersen, P., Holmqvist, B., and Voorhoeve, P. E. (1966b). Entorhinal activation of dentate granule cells. *Acta Physiol. Scand.* **66,** 448–460.

Axmacher, N., Mormann, F., Fernandez, G., Elger, C. E., and Fell, J. (2006). Memory formation by neuronal synchronization. *Brain Res. Rev.* **52,** 170–182.

Barnea, A., and Nottebohm, F. (1994). Seasonal recruitment of hippocampal neurons in adult free-ranging black-capped chickadees. *Proc. Natl. Acad. Sci. USA* **91,** 11217–11221.

Bayer, S. A. (1980a). Development of the hippocampal region in the rat. I. Neurogenesis examined with 3H-thymidine autoradiography. *J. Comp. Neurol.* **190,** 87–114.

Bayer, S. A. (1980b). Development of the hippocampal region in the rat. II. Morphogenesis during embryonic and early postnatal life. *J. Comp. Neurol.* **190,** 115–134.

Bayer, S. A., and Altman, J. (1987). Directions in neurogenetic gradients and patterns of anatomical connections in the telencephalon. *Prog. Neurobiol.* **29,** 57–106.

Becker, S. (2005). A computational principle for hippocampal learning and neurogenesis. *Hippocampus* **15,** 722–738.

Bizon, J. L., and Gallagher, M. (2003). Production of new cells in the rat dentate gyrus over the lifespan: Relation to cognitive decline. *Eur. J. Neurosci.* **18,** 215–219.

Bozon, B., Davis, S., and Laroche, S. (2002). Regulated transcription of the immediate-early gene Zif268: Mechanisms and gene dosage-dependent function in synaptic plasticity and memory formation. *Hippocampus* **12,** 570–577.

Breunig, J. J., Sarkisian, M. R., Arellano, J. I., Morozov, Y. M., Ayoub, A. E., Sojitra, S., Wang, B., Flavell, R. A., Rakic, P., and Town, T. (2008). Primary cilia regulate hippocampal neurogenesis by mediating Sonic hedgehog signaling. *Proc. Natl. Acad. Sci. USA* **105,** 13127–13132.

Brown, T. H., Kairiss, E. W., and Keenan, C. L. (1990). Hebbian synapses: Biophysical mechanisms and algorithms. *Annu. Rev. Neurosci.* **13,** 475–511.

Bruel-Jungerman, E., Laroche, S., and Rampon, C. (2005). New neurons in the dentate gyrus are involved in the expression of enhanced long-term memory following environmental enrichment. *Eur. J. Neurosci.* **21,** 513–521.

Bruel-Jungerman, E., Davis, S., Rampon, C., and Laroche, S. (2006). Long-term potentiation enhances neurogenesis in the adult dentate gyrus. *J. Neurosci.* **26,** 5888–5893.

Cameron, H. A., and McKay, R. D. (2001). Adult neurogenesis produces a large pool of new granule cells in the dentate gyrus. *J. Comp. Neurol.* **435,** 406–417.

Cao, L., Jiao, X., Zuzga, D. S., Liu, Y., Fong, D. M., Young, D., and During, M. J. (2004). VEGF links hippocampal activity with neurogenesis, learning and memory. *Nat. Genet.* **36,** 827–835.

Ceranik, K., Deng, J., Heimrich, B., Lubke, J., Zhao, S., Forster, E., and Frotscher, M. (1999). Hippocampal Cajal–Retzius cells project to the entorhinal cortex: Retrograde tracing and intracellular labelling studies. *Eur. J. Neurosci.* **11,** 4278–4290.

Chambers, R. A., Potenza, M. N., Hoffman, R. E., and Miranker, W. (2004). Simulated apoptosis/neurogenesis regulates learning and memory capabilities of adaptive neural networks. *Neuropsychopharmacology* **29,** 747–758.

Chawla, M. K., Guzowski, J. F., Ramirez-Amaya, V., Lipa, P., Hoffman, K. L., Marriott, L. K., Worley, P. F., McNaughton, B. L., and Barnes, C. A. (2005). Sparse, environmentally selective expression of Arc RNA in the upper blade of the rodent fascia dentata by brief spatial experience. *Hippocampus* **15,** 579–586.

Claiborne, B. J., Amaral, D. G., and Cowan, W. M. (1986). A light and electron microscopic analysis of the mossy fibers of the rat dentate gyrus. *J. Comp. Neurol.* **246,** 435–458.

Deisseroth, K., Singla, S., Toda, H., Monje, M., Palmer, T. D., and Malenka, R. C. (2004). Excitation-neurogenesis coupling in adult neural stem/progenitor cells. *Neuron* **42,** 535–552.

Del Rio, J. A., Heimrich, B., Borrell, V., Forster, E., Drakew, A., Alcantara, S., Nakajima, K., Miyata, T., Ogawa, M., Mikoshiba, K., Derer, P., Frotscher, M., *et al.* (1997). A role for Cajal–Retzius cells and reelin in the development of hippocampal connections. *Nature* **385,** 70–74.

Deller, T., Drakew, A., and Frotscher, M. (1999). Different primary target cells are important for fiber lamination in the fascia dentata: A lesson from reeler mutant mice. *Exp. Neurol.* **156,** 239–253.

Dobrossy, M. D., Drapeau, E., Aurousseau, C., Le Moal, M., Piazza, P. V., and Abrous, D. N. (2003). Differential effects of learning on neurogenesis: Learning increases or decreases the number of newly born cells depending on their birth date. *Mol. Psychiatry* **8,** 974–982.

Drapeau, E., and Nora Abrous, D. (2008). Role of neurogenesis in age-related memory disorders. *Aging Cell* **7,** 569–589.

Drapeau, E., Mayo, W., Aurousseau, C., Le Moal, M., Piazza, P. V., and Abrous, D. N. (2003). Spatial memory performances of aged rats in the water maze predict levels of hippocampal neurogenesis. *Proc. Natl. Acad. Sci. USA* **100,** 14385–14390.

Dupret, D., Montaron, M. F., Drapeau, E., Aurousseau, C., Le Moal, M., Piazza, P. V., and Abrous, D. N. (2005). Methylazoxymethanol acetate does not fully block cell genesis in the young and aged dentate gyrus. *Eur. J. Neurosci.* **22,** 778–783.

Dupret, D., Revest, J. M., Koehl, M., Ichas, F., De Giorgi, F., Costet, P., Abrous, D. N., and Piazza, P. V. (2008). Spatial relational memory requires hippocampal adult neurogenesis. *PLoS ONE* **3,** e1959.

Epp, J. R., Spritzer, M. D., and Galea, L. A. (2007). Hippocampus-dependent learning promotes survival of new neurons in the dentate gyrus at a specific time during cell maturation. *Neuroscience* **149,** 273–285.

Esposito, M. S., Piatti, V. C., Laplagne, D. A., Morgenstern, N. A., Ferrari, C. C., Pitossi, F. J., and Schinder, A. F. (2005). Neuronal differentiation in the adult hippocampus recapitulates embryonic development. *J. Neurosci.* **25,** 10074–10086.

Feng, R., Rampon, C., Tang, Y. P., Shrom, D., Jin, J., Kyin, M., Sopher, B., Miller, M. W., Ware, C. B., Martin, G. M., Kim, S. H., Langdon, R. B., *et al.* (2001). Deficient neurogenesis in forebrain-specific presenilin-1 knockout mice is associated with reduced clearance of hippocampal memory traces. *Neuron* **32,** 911–926.

Filippov, V., Kronenberg, G., Pivneva, T., Reuter, K., Steiner, B., Wang, L. P., Yamaguchi, M., Kettenmann, H., and Kempermann, G. (2003). Subpopulation of nestin-expressing progenitor cells in the adult murine hippocampus shows electrophysiological and morphological characteristics of astrocytes. *Mol. Cell. Neurosci.* **23,** 373–382.

Forster, E., Zhao, S., and Frotscher, M. (2006). Laminating the hippocampus. *Nat. Rev. Neurosci.* **7,** 259–267.

Frotscher, M. (1998). Cajal–Retzius cells, reelin, and the formation of layers. *Curr. Opin. Neurobiol.* **8,** 570–575.

Frotscher, M., Seress, L., Abraham, H., and Heimrich, B. (2001). Early generated Cajal–Retzius cells have different functions in cortical development. *Symp. Soc. Exp. Biol.* **53,** 43–49.

Fukuda, S., Kato, F., Tozuka, Y., Yamaguchi, M., Miyamoto, Y., and Hisatsune, T. (2003). Two distinct subpopulations of nestin-positive cells in adult mouse dentate gyrus. *J. Neurosci.* **23,** 9357–9366.

Gage, F. H. (2000). Mammalian neural stem cells. *Science* **287,** 1433–1438.

Garcia, A. D., Doan, N. B., Imura, T., Bush, T. G., and Sofroniew, M. V. (2004). GFAP-expressing progenitors are the principal source of constitutive neurogenesis in adult mouse forebrain. *Nat. Neurosci.* **7,** 1233–1241.

Ge, S., Goh, E. L., Sailor, K. A., Kitabatake, Y., Ming, G. L., and Song, H. (2006). GABA regulates synaptic integration of newly generated neurons in the adult brain. *Nature* **439,** 589–593.

Ge, S., Pradhan, D. A., Ming, G. L., and Song, H. (2007). GABA sets the tempo for activity-dependent adult neurogenesis. *Trends Neurosci.* **30,** 1–8.

Gould, E., Beylin, A., Tanapat, P., Reeves, A., and Shors, T. J. (1999). Learning enhance adult neurogenesis in the hippocampal formation. *Nat. Neurosci.* **2,** 260–265.

Gould, E., McEwen, B. S., Tanapat, P., Galea, L. A., and Fuchs, E. (1997). Neurogenesis in the dentate gyrus of the adult tree shrew is regulated by psychosocial stress and NMDA receptor activation. *J. Neurosci.* **17,** 2492–2498.

Gould, E., Tanapat, P., McEwen, B. S., Flugge, G., and Fuchs, E. (1998). Proliferation of granule cell precursors in the dentate gyrus of adult monkeys is diminished by stress. *Proc. Natl. Acad. Sci. USA* **95,** 3168–3171.

Guzowski, J. F., Lyford, G. L., Stevenson, G. D., Houston, F. P., McGaugh, J. L., Worley, P. F., and Barnes, C. A. (2000). Inhibition of activity-dependent arc protein expression in the rat hippocampus impairs the maintenance of long-term potentiation and the consolidation of long-term memory. *J. Neurosci.* **20,** 3993–4001.

Hairston, I. S., Little, M. T., Scanlon, M. D., Barakat, M. T., Palmer, T. D., Sapolsky, R. M., and Heller, H. C. (2005). Sleep restriction suppresses neurogenesis induced by hippocampus-dependent learning. *J. Neurophysiol.* **94,** 4224–4233.

Han, Y. G., Spassky, N., Romaguera-Ros, M., Garcia-Verdugo, J. M., Aguilar, A., Schneider-Maunoury, S., and Alvarez-Buylla, A. (2008). Hedgehog signaling and primary cilia are required for the formation of adult neural stem cells. *Nat. Neurosci.* **11,** 277–284.

Hastings, N. B., and Gould, E. (1999). Rapid extension of axons into the CA3 region by adult-generated granule cells. *J. Comp. Neurol.* **413,** 146–154.

Healy, S. D., Gwinner, E., and Krebs, J. R. (1996). Hippocampal volume in migratory and non-migratory warblers: Effects of age and experience. *Behav. Brain Res.* **81,** 61–68.

Jakubs, K., Nanobashvili, A., Bonde, S., Ekdahl, C. T., Kokaia, Z., Kokaia, M., and Lindvall, O. (2006). Environment matters: Synaptic properties of neurons born in the epileptic adult brain develop to reduce excitability. *Neuron* **52,** 1047–1059.

Karten, Y. J., Jones, M. A., Jeurling, S. I., and Cameron, H. A. (2006). GABAergic signaling in young granule cells in the adult rat and mouse dentate gyrus. *Hippocampus* **16,** 312–320.

Kee, N., Teixeira, C. M., Wang, A. H., and Frankland, P. W. (2007). Preferential incorporation of adult-generated granule cells into spatial memory networks in the dentate gyrus. *Nat. Neurosci.* **10,** 355–362.

Kempermann, G., and Gage, F. H. (2002). Genetic determinants of adult hippocampal neurogenesis correlate with acquisition, but not probe trial performance, in the water maze task. *Eur. J. Neurosci.* **16,** 129–136.
Kempermann, G., Kuhn, H. G., and Gage, F. H. (1997a). Genetic influence on neurogenesis in the dentate gyrus of adult mice. *Proc. Natl. Acad. Sci. USA* **94,** 10409–10414.
Kempermann, G., Kuhn, H. G., and Gage, F. H. (1997b). More hippocampal neurons in adult mice living in an enriched environment. *Nature* **386,** 493–495.
Kempermann, G., Brandon, E. P., and Gage, F. H. (1998a). Environmental stimulation of 129/SvJ mice causes increased cell proliferation and neurogenesis in the adult dentate gyrus. *Curr. Biol.* **8,** 939–942.
Kempermann, G., Kuhn, H. G., and Gage, F. H. (1998b). Experience-induced neurogenesis in the senescent dentate gyrus. *J. Neurosci.* **18,** 3206–3212.
Kempermann, G., Jessberger, S., Steiner, B., and Kronenberg, G. (2004). Milestones of neuronal development in the adult hippocampus. *Trends Neurosci.* **27,** 447–452.
Kohler, C. (1986). Intrinsic connections of the retrohippocampal region in the rat brain. II. The medial entorhinal area. *J. Comp. Neurol.* **246,** 149–169.
Laplagne, D. A., Esposito, M. S., Piatti, V. C., Morgenstern, N. A., Zhao, C., van Praag, H., Gage, F. H., and Schinder, A. F. (2006). Functional convergence of neurons generated in the developing and adult hippocampus. *PLoS Biol.* **4,** e409.
Laplagne, D. A., Kamienkowski, J. E., Esposito, M. S., Piatti, V. C., Zhao, C., Gage, F. H., and Schinder, A. F. (2007). Similar GABAergic inputs in dentate granule cells born during embryonic and adult neurogenesis. *Eur. J. Neurosci.* **25,** 2973–2981.
Lemaire, V., Koehl, M., Le Moal, M., and Abrous, D. N. (2000). Prenatal stress produces learning deficits associated with an inhibition of neurogenesis in the hippocampus. *Proc. Natl. Acad. Sci. USA* **97,** 11032–11037.
Leuner, B., Mendolia-Loffredo, S., Kozorovitskiy, Y., Samburg, D., Gould, E., and Shors, T. J. (2004). Learning enhances the survival of new neurons beyond the time when the hippocampus is required for memory. *J. Neurosci.* **24,** 7477–7481.
Liu, S., Wang, J., Zhu, D., Fu, Y., Lukowiak, K., and Lu, Y. M. (2003). Generation of functional inhibitory neurons in the adult rat hippocampus. *J. Neurosci.* **23,** 732–736.
Liu, X., Bolteus, A. J., Balkin, D. M., Henschel, O., and Bordey, A. (2006). GFAP-expressing cells in the postnatal subventricular zone display a unique glial phenotype intermediate between radial glia and astrocytes. *Glia* **54,** 394–410.
Lledo, P. M., Alonso, M., and Grubb, M. S. (2006). Adult neurogenesis and functional plasticity in neuronal circuits. *Nat. Rev. Neurosci.* **7,** 179–193.
Madsen, T. M., Kristjansen, P. E., Bolwig, T. G., and Wortwein, G. (2003). Arrested neuronal proliferation and impaired hippocampal function following fractionated brain irradiation in the adult rat. *Neuroscience* **119,** 635–642.
Meltzer, L. A., Yabaluri, R., and Deisseroth, K. (2005). A role for circuit homeostasis in adult neurogenesis. *Trends Neurosci.* **28,** 653–660.
Meshi, D., Drew, M. R., Saxe, M., Ansorge, M. S., David, D., Santarelli, L., Malapani, C., Moore, H., and Hen, R. (2006). Hippocampal neurogenesis is not required for behavioral effects of environmental enrichment. *Nat. Neurosci.* **9,** 729–731.
Mirescu, C., and Gould, E. (2006). Stress and adult neurogenesis. *Hippocampus* **16,** 233–238.
Mohapel, P., Leanza, G., Kokaia, M., and Lindvall, O. (2005). Forebrain acetylcholine regulates adult hippocampal neurogenesis and learning. *Neurobiol. Aging* **26,** 939–946.
Monje, M. L., Mizumatsu, S., Fike, J. R., and Palmer, T. D. (2002). Irradiation induces neural precursor-cell dysfunction. *Nat. Med.* **8,** 955–962.
Moser, E. I. (1996). Altered inhibition of dentate granule cells during spatial learning in an exploration task. *J. Neurosci.* **16,** 1247–1259.

Muller, A., Kukley, M., Stausberg, P., Beck, H., Muller, W., and Dietrich, D. (2005). Endogenous Ca^{2+} buffer concentration and Ca^{2+} microdomains in hippocampal neurons. *J. Neurosci.* **25,** 558–565.
Namba, T., Mochizuki, H., Onodera, M., Mizuno, Y., Namiki, H., and Seki, T. (2005). The fate of neural progenitor cells expressing astrocytic and radial glial markers in the postnatal rat dentate gyrus. *Eur. J. Neurosci.* **22,** 1928–1941.
Neves, G., Cooke, S. F., and Bliss, T. V. (2008). Synaptic plasticity, memory and the hippocampus: A neural network approach to causality. *Nat. Rev. Neurosci.* **9,** 65–75.
Nilsson, M., Perfilieva, E., Johansson, U., Orwar, O., and Eriksson, P. S. (1999). Enriched environment increases neurogenesis in the adult rat dentate gyrus and improves spatial memory. *J. Neurobiol.* **39,** 569–578.
Overstreet Wadiche, L., Bromberg, D. A., Bensen, A. L., and Westbrook, G. L. (2005). GABAergic signaling to newborn neurons in dentate gyrus. *J. Neurophysiol.* **94,** 4528–4532.
Overstreet-Wadiche, L. S., Bensen, A. L., and Westbrook, G. L. (2006). Delayed development of adult-generated granule cells in dentate gyrus. *J. Neurosci.* **26,** 2326–2334.
Raber, J., Rola, R., LeFevour, A., Morhardt, D., Curley, J., Mizumatsu, S., VandenBerg, S. R., and Fike, J. R. (2004). Radiation-induced cognitive impairments are associated with changes in indicators of hippocampal neurogenesis. *Radiat. Res.* **162,** 39–47.
Ramirez-Amaya, V., Marrone, D. F., Gage, F. H., Worley, P. F., and Barnes, C. A. (2006). Integration of new neurons into functional neural networks. *J. Neurosci.* **26,** 12237–12241.
Santarelli, L., Saxe, M., Gross, C., Surget, A., Battaglia, F., Dulawa, S., Weisstaub, N., Lee, J., Duman, R., Arancio, O., Belzung, C., and Hen, R. (2003). Requirement of hippocampal neurogenesis for the behavioral effects of antidepressants. *Science* **301,** 805–809.
Saxe, M. D., Battaglia, F., Wang, J. W., Malleret, G., David, D. J., Monckton, J. E., Garcia, A. D., Sofroniew, M. V., Kandel, E. R., Santarelli, L., Hen, R., and Drew, M. R. (2006). Ablation of hippocampal neurogenesis impairs contextual fear conditioning and synaptic plasticity in the dentate gyrus. *Proc. Natl. Acad. Sci. USA* **103,** 17501–17506.
Schmidt-Hieber, C., Jonas, P., and Bischofberger, J. (2004). Enhanced synaptic plasticity in newly generated granule cells of the adult hippocampus. *Nature* **429,** 184–187.
Seri, B., Garcia-Verdugo, J. M., Collado-Morente, L., McEwen, B. S., and Alvarez-Buylla, A. (2004). Cell types, lineage, and architecture of the germinal zone in the adult dentate gyrus. *J. Comp. Neurol.* **478,** 359–378.
Sherry, D. F., Forbes, M. R., Khurgel, M., and Ivy, G. O. (1993). Females have a larger hippocampus than males in the brood-parasitic brown-headed cowbird. *Proc. Natl. Acad. Sci. USA* **90,** 7839–7843.
Shors, T. J., Miesegaes, G., Beylin, A., Zhao, M., Rydel, T., and Gould, E. (2001). Neurogenesis in the adult is involved in the formation of trace memories. *Nature* **410,** 372–376.
Shors, T. J., Townsend, D. A., Zhao, M., Kozorovitskiy, Y., and Gould, E. (2002). Neurogenesis may relate to some but not all types of hippocampal-dependent learning. *Hippocampus* **12,** 578–584.
Snyder, J. S., Kee, N., and Wojtowicz, J. M. (2001). Effects of adult neurogenesis on synaptic plasticity in the rat dentate gyrus. *J. Neurophysiol.* **85,** 2423–2431.
Snyder, J. S., Hong, N. S., McDonald, R. J., and Wojtowicz, J. M. (2005). A role for adult neurogenesis in spatial long-term memory. *Neuroscience* **130,** 843–852.
Song, H. J., Stevens, C. F., and Gage, F. H. (2002). Neural stem cells from adult hippocampus develop essential properties of functional CNS neurons. *Nat. Neurosci.* **5,** 438–445.

Squire, L. (1987). "Memory and Brain." Oxford University Press, New York.

Steiner, B., Klempin, F., Wang, L., Kott, M., Kettenmann, H., and Kempermann, G. (2006). Type-2 cells as link between glial and neuronal lineage in adult hippocampal neurogenesis. *Glia* **54,** 805–814.

Steward, O. (1976). Topographic organization of the projections from the entorhinal area to the hippocampal formation of the rat. *J. Comp. Neurol.* **167,** 285–314.

Suh, H., Consiglio, A., Ray, J., Sawai, T., D'Amour, K. A., and Gage, F. H. (2007). *In vivo* fate analysis reveals the multipotent and self-renewal capacities of Sox2(+) neural stem cells in the adult hippocampus. *Cell Stem Cell* **1,** 515–528.

Sundholm-Peters, N. L., Yang, H. K., Goings, G. E., Walker, A. S., and Szele, F. G. (2004). Radial glia-like cells at the base of the lateral ventricles in adult mice. *J. Neurocytol.* **33,** 153–164.

Super, H., and Soriano, E. (1994). The organization of the embryonic and early postnatal murine hippocampus. II. Development of entorhinal, commissural, and septal connections studied with the lipophilic tracer DiI. *J. Comp. Neurol.* **344,** 101–120.

Swanson, L. W. (1983). The hippocampus and concept of the limbic system. *In* "Neurobiology of the Hippocampus" (W. Seifert, ed.), pp. 3–19. Academic Press, London.

Tanapat, P., Hastings, N. B., Rydel, T. A., Galea, L. A., and Gould, E. (2001). Exposure to fox odor inhibits cell proliferation in the hippocampus of adult rats via an adrenal hormone-dependent mechanism. *J. Comp. Neurol.* **437,** 496–504.

Tashiro, A., Sandler, V. M., Toni, N., Zhao, C., and Gage, F. H. (2006). NMDA-receptor-mediated, cell-specific integration of new neurons in adult dentate gyrus. *Nature* **442,** 929–933.

Toni, N., Teng, E. M., Bushong, E. A., Aimone, J. B., Zhao, C., Consiglio, A., van Praag, H., Martone, M. E., Ellisman, M. H., and Gage, F. H. (2007). Synapse formation on neurons born in the adult hippocampus. *Nat. Neurosci.* **10,** 727–734.

Toni, N., Laplagne, D. A., Zhao, C., Lombardi, G., Ribak, C. E., Gage, F. H., and Schinder, A. F. (2008). Neurons born in the adult dentate gyrus form functional synapses with target cells. *Nat. Neurosci.* **11,** 901–907.

Tozuka, Y., Fukuda, S., Namba, T., Seki, T., and Hisatsune, T. (2005). GABAergic excitation promotes neuronal differentiation in adult hippocampal progenitor cells. *Neuron* **47,** 803–815.

van Praag, H., Kempermann, G., and Gage, F. H. (1999). Running increases cell proliferation and neurogenesis in the adult mouse dentate gyrus. *Nat. Neurosci.* **2,** 266–270.

van Praag, H., Schinder, A. F., Christie, B. R., Toni, N., Palmer, T. D., and Gage, F. H. (2002). Functional neurogenesis in the adult hippocampus. *Nature* **415,** 1030–1034.

van Praag, H., Shubert, T., Zhao, C., and Gage, F. H. (2005). Exercise enhances learning and hippocampal neurogenesis in aged mice. *J. Neurosci.* **25,** 8680–8685.

Wang, S., Scott, B. W., and Wojtowicz, J. M. (2000). Heterogeneous properties of dentate granule neurons in the adult rat. *J. Neurobiol.* **42,** 248–257.

Wang, L. P., Kempermann, G., and Kettenmann, H. (2005). A subpopulation of precursor cells in the mouse dentate gyrus receives synaptic GABAergic input. *Mol. Cell. Neurosci.* **29,** 181–189.

Winocur, G., Wojtowicz, J. M., Sekeres, M., Snyder, J. S., and Wang, S. (2006). Inhibition of neurogenesis interferes with hippocampus-dependent memory function. *Hippocampus* **16,** 296–304.

Wiskott, L., Rasch, M. J., and Kempermann, G. (2006). A functional hypothesis for adult hippocampal neurogenesis: Avoidance of catastrophic interference in the dentate gyrus. *Hippocampus* **16,** 329–343.

Zhang, F., Wang, L. P., Boyden, E. S., and Deisseroth, K. (2006). Channelrhodopsin-2 and optical control of excitable cells. *Nat. Methods* **3,** 785–792.

Zhang, C. L., Zou, Y., He, W., Gage, F. H., and Evans, R. M. (2008). A role for adult TLX-positive neural stem cells in learning and behaviour. *Nature* **451,** 1004–1007.

Zhao, S., Forster, E., Chai, X., and Frotscher, M. (2003a). Different signals control laminar specificity of commissural and entorhinal fibers to the dentate gyrus. *J. Neurosci.* **23,** 7351–7357.

Zhao, X., Ueba, T., Christie, B. R., Barkho, B., McConnell, M. J., Nakashima, K., Lein, E. S., Eadie, B. D., Willhoite, A. R., Muotri, A. R., Summers, R. G., Chun, J., *et al.* (2003b). Mice lacking methyl-CpG binding protein 1 have deficits in adult neurogenesis and hippocampal function. *Proc. Natl. Acad. Sci. USA* **100,** 6777–6782.

Zhao, C., Teng, E. M., Summers, R. G., Ming, G. L., and Gage, F. H. (2006). Distinct morphological stages of dentate granule neuron maturation in the adult mouse hippocampus. *J. Neurosci.* **26,** 3–11.

Zhao, C., Deng, W., and Gage, F. H. (2008). Mechanisms and functional implications of adult neurogenesis. *Cell* **132,** 645–660.

CHAPTER SIX

Looking Beyond Development: Maintaining Nervous System Architecture

Claire Bénard *and* Oliver Hobert

Contents

Abstract

Neuronal circuitries established in development must persist throughout life. This poses a serious challenge to the structural integrity of an embryonically patterned nervous system as an animal dramatically increases its size postnatally, remodels parts of its anatomy, and incorporates new neurons. In addition, body movements, injury, and ageing generate physical stress on the nervous system. Specific molecular pathways maintain intrinsic properties of neurons in the mature nervous system. Other factors ensure that the overall organization of entire neuronal ensembles into ganglia and fascicles is appropriately maintained upon external challenges. Here, we discuss different molecules underlying these neuronal maintenance mechanisms, with a focus on lessons learned from the nematode *Caenorhabditis elegans*.

Department of Biochemistry and Molecular Biophysics, Howard Hughes Medical Institute, Columbia University Medical Center, New York, USA

Current Topics in Developmental Biology, Volume 87
ISSN 0070-2153, DOI: 10.1016/S0070-2153(09)01206-X

1. Introduction

1.1. The maturing brain

A series of well-orchestrated processes build our brain *in utero*. The neural tube forms by 3–4 weeks of gestation, differentiates into regions, and hosts massive neuronal proliferation and migration during weeks 12–20. Synapses then organize, while extensive cell death sculpts the forming brain until the first few weeks of postnatal life. At birth, our brain is made of approximately 100 billion neurons, but a dynamic and lengthy process of maturation of the brain ensues to increase this number further (Fig. 6.1). Profound modifications of the anatomy occur on a large scale, involving growth of the entire brain and some subregions, concomitant with widespread changes at the scale of the neuron, its axon, dendrites, and synapses. The organization of the brain established by the time of birth must be maintained throughout this process of maturation.

In parallel to this expansion of the brain, extensive changes of cortical and subcortical components occur during childhood and adolescence

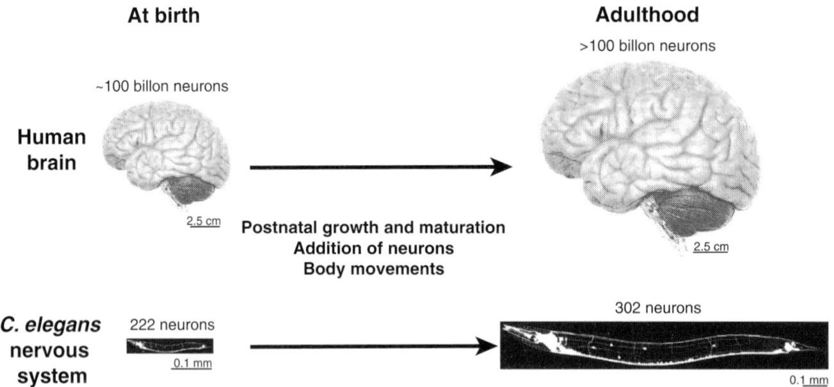

Figure 6.1 After the nervous system is established during embryogenesis, it faces the challenges of growth, maturation, and mechanical strain. At birth, a human brain is only about 1/4–1/3 of our adult brain volume. It grows and develops most dramatically during childhood. The fastest growth occurs in the first 3 years of postnatal life, when the infant's brain reaches 90% of the adult weight. By age 6, the total brain size is 95% of its maximum, reaching the adult volume and weight by about 12 years of age (Lenroot and Giedd, 2006; Toga *et al.*, 2006). Although not shown in the diagram, other regions of the central and peripheral nervous systems also grow expansively during childhood and adolescence. The nervous system of the worm *C. elegans* (bottom) is laid out in embryogenesis. After hatching into a larva (birth), the worm grows 100 times in volume, and the entire nervous system grows considerably as well. The nervous system of both organisms are subjected to mechanical strain originating from growth itself, the addition of neurons and their axons, and the motions of the body.

(Lenroot and Giedd, 2006). In the first few years of life, dendritic branching and synaptic connections increase tremendously, followed by dendritic pruning and synapse elimination, which result in the selection of the most efficient and active connections. In particular, synaptogenesis peaks at 4 months of postnatal life in the visual cortex, and synapse elimination brings about the adult state by 4–5 years of age. In the medial prefrontal cortex, synaptic growth peaks at 3–4 years and the bulk of trimming is completed by mid-to-late adolescence (Toga *et al.*, 2006). Changes in volume and density of gray and white matter continue well into the third decade of life. Vast waves of myelination take place during late adolescence and continue in the adult.

Taken together, brain development and maturation are highly dynamic, continuous processes that involve substantial growth and morphological alteration during embryonic and postembryonic stages, respectively. Moreover, the maturing brain is endowed with great plasticity that is essential for shaping its structure and function. Plasticity is most prevalent in defined time windows or critical periods (Hubel and Wiesel, 1970), but structural changes continue to occur. For example, the adult cortex undergoes structural changes in spines (Trachtenberg *et al.*, 2002), axons (Florence *et al.*, 1998), and dendrites (Tailby *et al.*, 2005). Also, adult neurogenesis generates a substantial number of new neurons that are incorporated into existing circuitries (Li, Mu, and Gage, this volume). In spite of these massive alterations, many of the neuronal features and structures established earlier in development are retained throughout life.

1.2. Challenges faced by nervous systems during maturation

The dramatic changes that occur during brain maturation raise the question of how neuroanatomical structures formed at an earlier time point retain their essential morphology and connectivity amidst the expansive growth of the nervous system and the remodeling of many of its connections, and beyond. The overall growth of the human brain by 3–4 times during childhood constitutes a challenge for the existing neuroanatomical structures as substantial tension is generated on whole brain regions, neurons, and neighboring cells. Tension is produced at the level of entire organs undergoing growth, and likely is engendered by the massive volume increase during brain maturation (Van Essen, 1997). Also, the growth of the body implies that individual neurons grow as well if they are to remain connected to their distant targets (e.g., as the limbs elongate, axons of the sciatic nerves grow such that they still reach the toe).

The addition of new cells, their growth, and migration also exert mechanical forces on neurons close by, albeit more locally. Also, the addition of new neurons, cells, and synaptic arbors profoundly alters the molecular landscape of the neurons that are acquiring new neighbors. The ensheathing of axons by Schwann cells or oligodendrocytes replaces prior contacts between neighbors and establishes new ones, with different cell surface molecules. In the face of the mechanical and molecular pressures generated during brain maturation, at least some neuroanatomical features that developed earlier likely need to remain connected, and accommodate and integrate the novelty around them.

1.3. Challenges faced by nervous systems throughout life

The nervous system not only faces challenges of various sorts during growth and maturation, but also throughout adult life. This is because nervous systems are not fixed in a static state and remain plastic throughout adulthood, incorporating activity-dependent changes and remodeling, refining, strengthening, adding, and eliminating synapses (Sanes and Lichtman, 2001). In addition, the movements of our body parts are a source of ongoing, lifelong mechanical stress on neuronal structures. For instance, neurons of the peripheral nervous system (PNS) are subjected to tension by the actions of our hands, limbs, eyes, tongue, etc. Also, within the central nervous system (CNS), a structure like the optic nerve is under continuous mechanical stress imposed by eye movements. Thus, the mechanical and molecular challenges, local and widespread, loom throughout life. This demands a finely orchestrated balance between flexibility, and plasticity on the one hand, and the stability and maintenance of structures, on the other.

1.4. Levels of neuronal maintenance

There are two different aspects to the maintenance of structure and function of the nervous system. First, like any other cell type, neuronal cells maintain intrinsic structural and functional features. Second, on a broader scale, the overall organization of ensembles of neurons is maintained, which involves the maintenance of the correct position of neurons (soma, axon, dendrites) relative to one another and relative to other components of the nervous system, such as glial cells. Such maintenance is expected to be absolutely critical to ensure that neurons retain their ability to appropriately communicate with one another throughout periods of postnatal growth, remodeling and mechanical stress.

2. Maintenance of Cellular Features of Neurons

The most obvious and fundamental cellular aspects that neurons must maintain are (a) their gene expression programs, which define their identity and functional properties, and (b) their cellular survival. The maintenance of gene expression programs involves the continuous activity of transcription factors, often those that at first initiate neuron-type-specific transcriptional events, but also chromatin factors that serve to maintain activated and repressed states of transcription (Brock and Fisher, 2005; Shirasaki and Pfaff, 2002). Neuronal survival is ensured through a host of well-known trophic factors such as neurotrophins (Henderson, 1996).

Moreover, the integrity and morphology of a neuron and its subcellular parts, including axons, dendrites, and synapses, require active maintenance. This is well illustrated by genetic mutant analysis, which revealed that the removal of specific genes does not affect the initial generation of specific neuronal features, but affects their continuous maintenance. A few examples include the following. In flies, the genes *warts* and *hippo* are required for maintenance of dendritic tiling (Emoto *et al.*, 2006). *warts* mutants initially have their body wall normally tiled by dendritic branches. However, from several days post-egg laying on, dendrites progressively lose branches. Dendritic arbor maintenance is also defective in mouse knockouts of the Abelson family kinases. In these mutants, dendritic arbor is normal in immature mice, but defects unfold in adult cortical neurons, only after 3 weeks of postnatal life, that is, just after the most intensive period of dendrite growth in the mouse cortex (Moresco *et al.*, 2005). Illustrating the disease relevance of neuronal maintenance, cortical neurons of Down's syndrome patients gradually lose dendritic branches after initially forming fully branched dendritic fields (Kaufmann and Moser, 2000). Neurological symptoms appear only late in development, after dendrite structure is already established, suggesting that dendrites deteriorate. Indeed, dendritic length is normal or exceedingly long in fetuses and infants with Down's syndrome, but steadily decreases in subjects older than 2 years. Even though it is not quite clear which specific molecule is involved in this Down's syndrome phenotype, there are several other molecules known to be involved in maintaining dendritic complexity, such as BDNF and TrkB (Gorski *et al.*, 2003; Xu *et al.*, 2000), integrins (Marrs *et al.*, 2006; Peng *et al.*, 2008), and collagen IV (Fox *et al.*, 2007).

The need for specific mechanisms to maintain axonal integrity is well illustrated by several examples. For instance, axonal transport is crucial in maintaining the axon scaffold. Postnatal disruption of the dynein/dynactin complex that inhibits retrograde axonal transport leads to late-onset axon degeneration in mice (LaMonte *et al.*, 2002). Neuronal polarity also appears to be actively maintained. After having grown a single axon and neurites in culture, neurons convert pre-existing dendrites into axons, generating multiple axons, as a result of disrupted function of members of the serine/threonine protein kinase D that disturbs membrane trafficking in the Golgi apparatus (Yin *et al.*, 2008).

Synapses also require active maintenance, as revealed by the analysis of genes whose depletion causes a progressive loss of synaptic contacts. This has been best illustrated at the neuromuscular junction, as reviewed earlier

(Sanes and Lichtman, 2001). More recently, a series of distinct target-derived signals has been identified that organize the formation, maturation, and maintenance of motor nerve terminals. In particular, synapse-specific collagen IV chains (alpha 3–6) accumulate only after synapses are mature and are required for synaptic maintenance (Fox *et al.*, 2007).

A recurrent theme in maintaining the morphology of individual neurons is the activity-dependent suppression of sprouting of individual neuronal processes (Cao *et al.*, 2007; Huberman *et al.*, 2008). Classic drug studies have demonstrated that neurons alter their morphology in response to inhibition of synaptic transmission, resulting in the growth of sprouts, an apparent attempt of a neuron to re-establish a functional connection (Brown *et al.*, 1981). Genetic analysis has amply illustrated this point further. For example, mutations in the cyclic nucleotide-gated channel subunits *tax-2* and *tax-4* in *C. elegans*, which are required for sensory neuron signal transduction, lead to abnormal axonal branching or the failure of axons to terminate at normal positions (Coburn *et al.*, 1998; Peckol *et al.*, 1999). The time of action for these genes is strictly postdevelopmental, demonstrating the continuous need of these molecules to maintain normal structural features of a neuron (Coburn *et al.*, 1998). Postdevelopmental, supernumerary axonal branching defects can also be observed upon genetic disruption of trans-synaptic communication (Loria *et al.*, 2004; Zhao and Nonet, 2000). These observations point to as yet poorly understood retrograde feedback signals that monitor synaptic activity and signal toward morphological rearrangements upon loss of appropriate synaptic activity. Other molecules involved in maintenance of structural properties include, for example, the *unc-119* locus in *C. elegans* (Knobel *et al.*, 2001). The UNC-119 protein is located in axons and functions in neurons throughout larval stages, after neurons completed developing their morphology, where it inhibits excessive axon branching.

A completely distinct example of a molecule that maintains intrinsic features of a neuron is the *Drosophila* protein Spam. Spam forms an extracellular shield that protects mechanosensory neurons from massive cellular deformation caused by heat-induced osmotic imbalance (Cook *et al.*, 2008). This example illustrates the importance for a neuron to utilize specific molecules to cope with postnatal challenges.

3. Maintenance of the Architecture of Neuronal Ensembles

3.1. Neuron/glia interactions

Molecular mechanisms that support the maintenance of synaptic contacts suggest that a neuron must maintain not only its intrinsic cellular properties, but also the interactions with its cellular environment. In the vertebrate nervous system, one context in which this is very clear is that of neuron/glia interactions. A major function of glia in the adult is the formation of myelin sheaths around axons thus allowing the fast conduction of signaling essential for nervous system function. Glia also maintain appropriate concentrations of ions and neurotransmitters in the neuronal environment (Jessen, 2004). Studies of neuron/glia interactions have revealed ample evidence for the involvement of specific molecules in not only generating but also maintaining specific cellular contacts and we briefly mention a few examples below.

Oligodendrocytes and Schwann cells are critically involved in the long-term survival of axons in the CNS and PNS, respectively. Two myelin membrane proteolipids, PLP and DM20, are essential for the integrity of myelinated axons. In the absence of PLP–DM20, mice assemble compact myelin sheaths but subsequently develop widespread axonal swellings and degeneration. From the age of 6 to 8 weeks, focal axonal swellings containing organelles are detected throughout the white and gray matter in all regions of the CNS. By 1 year of age, numerous axonal swellings in the optic nerve and spinal cord occur, and at older ages, the motor performance of mutant mice is altered (Griffiths *et al.*, 1998). Similarly, PLP-null mutants have only subtle myelin defects and are neurologically normal in the first year of life. At older ages, a destabilization of compacted myelin arises (Klugmann *et al.*, 1997). Mice deficient in the gene for myelin-associated glycoprotein (MAG) also develop normal myelin sheaths in the PNS, but in mutant mice older than 8 months the maintenance of axon–myelin units is disturbed, resulting in both axon and myelin degeneration. MAG thus appears to play a crucial role in the long-term maintenance of the integrity of both myelin and axons (Fruttiger *et al.*, 1995).

The examples from tightly associated glial cells are among the best case studies for the existence of maintenance factors that keep the overall organization of the nervous system intact. Mutant analysis in simple invertebrate nervous systems, where the contribution of glial cells is less prevalent that in the vertebrate nervous system, has substantially extended our view of the cellular interactions among neurons, as well as between neurons and surrounding tissue, to ensure the proper maintenance of whole neuronal ensembles, as we discuss in the following section.

3.2. Maintaining overall nervous system architecture

Beyond the maintenance of neurons at the individual cell level or at the level of interactions with its immediate surroundings and synaptic partners, architectural aspects of the nervous system must also be maintained. By architecture we mean the spatial arrangements of multineuronal assemblages that are organized into defined neuronal fascicles, commissures, and ganglia.

This definition not only encompasses defined neuronal circuits, but the overall arrangements of circuits into suprastructural ensembles.

A priori, it would be reasonable that maintenance of such structural organization simply relies on the same cues that have initially laid down the overall organization of the nervous system. However, genetic analysis in the nematode *C. elegans* has demonstrated that dedicated mechanisms exist to maintain neuronal architecture. These mechanisms ensure that axons within fascicles and neuronal soma within ganglia retain their precise position after initially adopting their appropriate position through developmental patterning mechanisms. *C. elegans* has been a particularly well-suited organism to identify such maintenance mechanisms as the normal structure of the nervous system is known in exquisite detail, its constituents (neuronal processes, neuronal cell bodies) can be visualized with single cell resolution at defined time points, and it is easily amenable to genetic analysis. Below, we will first very briefly describe the structure and development of the *C. elegans* nervous system and then walk through individual maintenance factors.

The nervous system of the adult *C. elegans* hermaphrodite is composed of 302 neurons. They are organized into well-defined ganglia and fascicles that are apposed to specific cell types such as epidermis and muscle, from which they are often separated by dense extracellular matrix material (Fig. 6.2) (White *et al.*, 1986). The most prominent feature of the worm's nervous system is its densely clustered head ganglia ("brain"), and its ventral nerve cord ("spinal cord"), which contains motor- and interneurons and is separated by a ventral midline. Axons are not myelinated and establish chemical and electrical *en passant* synapses along the length of the axons. Most (~75%) of the adult nervous system develops embryonically, both in terms of neuron birth, axon outgrowth and synaptic connectivity, but additional neurons, mostly ventral cord motor neurons, are added during the first larval stage.

The nervous system of *C. elegans* faces the same challenges as any other nervous system, in that (a) it undergoes enormous growth (Fig. 6.1), (b) preformed circuits must tolerate and integrate the addition of new neurons throughout postembryonic, early larval stages (Fig. 6.2), and (c) it must withstand mechanical stress (Fig. 6.2). Mechanical stress results from locomotion of the worm itself, first inside the eggshell and then after hatching. Mechanical stress is also generated through the movement of internal organs, such as the pharynx, whose peristaltic pumping movements exert pressure on neighboring neurons in the main head ganglia (Fig. 6.2). Egg laying and defecation represent other repeated strains on the nervous system. In terms of growth, the worm increases in length and diameter during development, so that the body volume of an adult is 100 times larger than that of a hatchling at the end of embryogenesis, when its nervous system architecture has been laid out. This growth entails the lengthening of axons and dendrites of the embryonically born neurons.

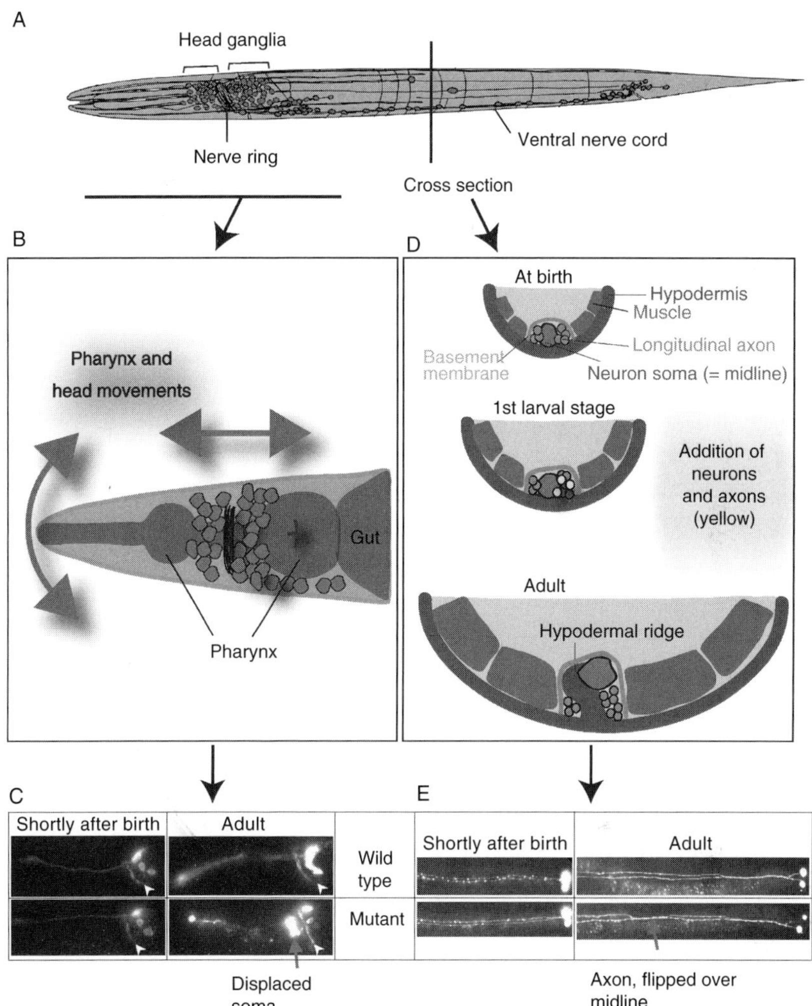

Figure 6.2 Maintenance of neuronal architecture in the nervous system of *C. elegans*. (A) The majority of the 302 neurons (in red) of the worm are organized into major ganglia in the head and tail, as well as into fascicles such as the ventral nerve cord. This organization is established during embryogenesis and retained throughout life. (B) Schematic representation of the head of a worm, with neurons in head ganglia (in red). Throughout larval and adult stages, the worm swims in its environment foraging for food and feeds through the motions of its pharynx, which pumps food toward the gut. These lifelong motions exert considerable mechanical strain on the neurons situated near the pharynx and lead to the progressive displacement of neurons in the absence of dedicated neuronal maintenance mechanisms (C). (C) Example of maintenance defects in head ganglia. In wild-type worms, the cell bodies of a subset of chemosensory neurons of the head are located posterior to the nerve ring (indicated by a white arrowhead). This chemosensory architecture develops in embryogenesis and persists throughout life. In maintenance mutants such as *sax-7* or *dig-1*, the chemosensory neurons develop

3.3. Molecules mediating neuronal maintenance in *C. elegans*

A combination of screens for mutants with defects in neuronal morphology and candidate gene approaches have revealed a number of genes that maintain nervous system architecture (Fig. 6.3). We term these genes "maintenance factors." Mutant phenotypes of the maintenance factors fall into essentially two categories. In one category, the position of axons within fascicles fails to be maintained, while in another category the relative position of neuronal soma within ganglia fails to be maintained (Fig. 6.2). Some mutants only affect axons, others only soma, while yet others affect both. Some of the molecules have developmental roles in other cellular contexts. Nevertheless in all cases, the defects observed in the respective mutants are strictly postdevelopmental. That is to say that mutant animals look indistinguishable from wild-type after the development of the neuron in question has terminated (having extended its axon, dendrites, branches, etc.), and only display defects in later stages of life. This suggests that these molecules are specifically involved in *maintaining* neuronal organization, rather than acting as developmental factors with continuous activity throughout postdevelopmental stages, as many of the cases discussed in the previous section. Lastly, maintenance defects in these mutant backgrounds can be suppressed by inhibiting locomotion (see below), suggesting that the defects are induced

---

completely normally and are indistinguishable form wild-type at birth, but later become progressively displaced (red arrow, anterior to the nerve ring). (D) Diagram of the ventral portion of a cross section through a worm at birth, first larval stage, and adulthood. The ventral nerve cord develops largely embryonically, with axons organized into the left and right fascicles on either side of a midline; the midline is initially constituted of motor neuron soma that are aligned along the *a/p*-axis (Boulin *et al.*, 2006) and post-embryonically becomes further elaborated by an evagination ("hypodermal ridge") of epidermal tissue. The ventral nerve cord is ensheathed by basement membrane material (shown in blue). This separate organization of the fascicles of the ventral nerve cord is maintained throughout life, despite the addition of neurons and axons during the first larval stage (in yellow). However, this arrangement can be disturbed, in the absence of dedicated maintenance factors, as a result of the mechanical strain exerted by the locomotion movements of the worm (E); the failure of maintaining ventral nerve cord architecture occurs at a time right after birth (first larval stage) when additional axons and neurons are added and when the hypodermal ridge—an insurmountable obstacle in adult animals—has not yet been fully elaborated. (E) Example of maintenance defects in the ventral nerve cord. In wild-type worms, the axons of the two PVQ neurons project into the left and the right fascicles of the ventral nerve cord, during embryogenesis. These axons remain in their precise position within the ventral nerve cord throughout life, despite incessant movements of locomotion. In *zig-4*, *egl-15*, *sax-7*, and *dig-1* mutants, these axons develop normally; however, during the first larval stage, mechanical strain from locomotion leads specific axons to flip over the ventral midline to the other fascicle of the ventral nerve cord. Small white dots are background autofluorescence from the gut. (See Color Insert.)

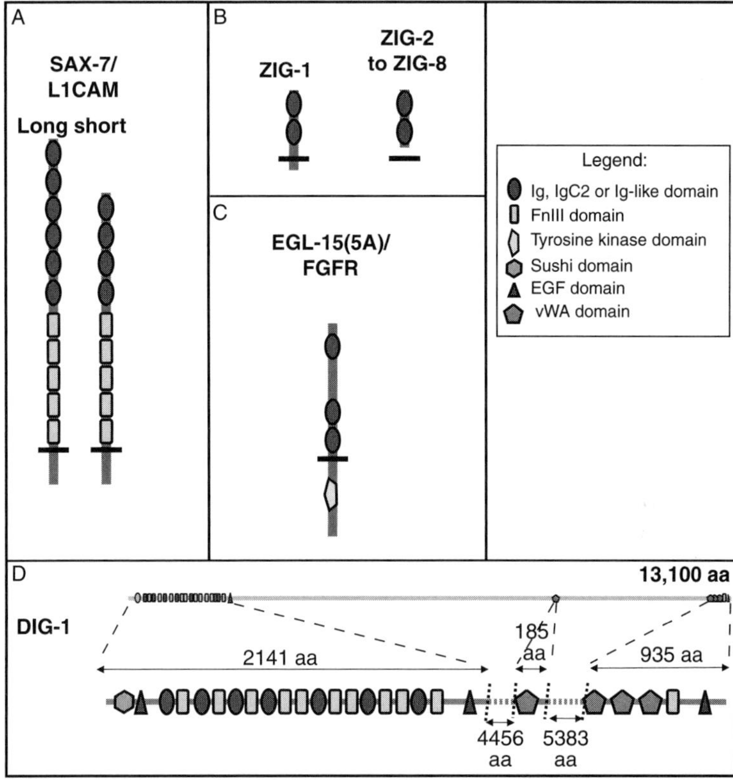

Figure 6.3 Molecules implicated in maintenance of neuronal architecture. To date, diverse members of the immunoglobulin superfamily are implicated in maintaining the precise position of axons within fascicles and neuronal soma in ganglia in *C. elegans*. (See Color Insert.)

by mechanical stress and that the normal function of the maintenance factors are to counteract such mechanical impacts. We describe the individual maintenance factors in more detail below.

3.3.1. SAX-7: A human disease gene ortholog of the IgCAM family

Animals carrying loss-of-function mutations in the *sax-7* gene display a large spectrum of neuronal maintenance defects, including a failure to maintain axon positioning in the ventral nerve cord, as well as the relative position of neuronal soma in various head ganglia and ventral nerve cord (Axang *et al.*, 2007; Chen *et al.*, 2001; Pocock *et al.*, 2008; Sasakura *et al.*, 2005; Wang *et al.*, 2005; Zallen *et al.*, 1999) (Fig. 6.2). These defects are only apparent postembryonically, that is, the nervous system initially develops normally in the embryo, but embryonically patterned neuronal structures lose their normal positioning after birth (hatching from the eggshell). These defects are partly suppressed if the worms are paralyzed, suggesting that movements lead to displacement, and that the wild-type function of *sax-7* is to antagonize the mechanical force triggered by locomotion (Sasakura *et al.*, 2005) and also, in the case of head neurons, peristaltic pumping of the adjacent pharynx.

sax-7 encodes a broadly expressed homophilic cell adhesion molecule with several extracellular immunoglobulin (Ig) and fibronectin type III (FnIII) domains (Fig. 6.3), which acts autonomously in the nervous system (Chen *et al.*, 2001; Pocock *et al.*, 2008; Sasakura *et al.*, 2005). The intracellular domain of SAX-7 directly interacts with the cytoskeleton and this interaction is important for neuronal maintenance (Zhou *et al.*, 2008). SAX-7 is one of the two orthologues of Vertebrate L1 family members (L1, CHL1, Neurofascin, and NrCAM), which have been implicated in many aspects of nervous system function and development (Rougon and Hobert, 2003). Mutations in human L1 result in a wide spectrum of neurological abnormalities, including X-linked hydrocephalus, MASA syndrome, X-linked complicated spastic paraplegia type 1, and X-linked agenesis of the corpus callosum (Fransen *et al.*, 1997). It is not clear whether these defects are developmental, as commonly assumed, or whether they are rather a reflection of maintenance defects. It is known, however, that the specific depletion of L1 in the adult mouse brain results in behavioral abnormalities, highlighting its functional importance in the adult brain (Law *et al.*, 2003). Although no gross anatomical abnormalities have been observed in these mice, it is possible that finer defects in the relative position of neurons exist.

3.3.2. DIG-1: A giant multidomain extracellular matrix protein

dig-1 mutants display phenotypes similar to those observed in *sax-7* mutants, namely a failure to maintain axonal position within fascicles and relative position of neuronal soma in head ganglia. A large number of neuronal soma and axonal projections acquire their proper position during development, but become displaced after birth of the animal (Benard *et al.*, 2006; Zallen *et al.*, 1999). Defects are suppressible by inhibition of movement, indicating that like *sax-7*, the *dig-1* gene serves to counteract mechanical stress exerted on neuronal structures (Benard *et al.*, 2006). The neuronal defects of *dig-1* mutants are not a consequence of gross developmental or morphological defects of neighboring muscles or epidermis, as these tissues appear completely normal. Also, the maintenance defects of axon and soma for a given neuron are not correlated.

DIG-1 is a huge secreted protein of 13,100 amino acids and contains numerous conserved domains implicated in extracellular protein interactions (Fig. 6.3). In its N-terminal region, DIG-1 contains Ig, FnIII, Sushi, and EGF domains, which are frequently part of proteins implicated in cell adhesion, and in components of the extracellular matrix. An enormous central region is composed of a large number of repeats, rich in β-strands that may form individual, globular domains. This central repetitive region harbors many Ser–Gly peptide motifs in an acidic context, which could be glycosaminoglycans attachment sites (Burket *et al.*, 2006; Lindahl and Hook, 1978). In the C-terminal region, DIG-1 contains Ig, EGF, and several von Willebrand factor A (vWA) domains, again commonly involved in cell adhesion and cell–cell interactions. This overall arrangement of domains and motifs in the DIG-1 protein make it akin to the hyalectan class of proteoglycans, with N- and C-terminal Ig, EGF, and Sushi/CCP domains and a central glycosaminoglycan attachment region (Bandtlow and Zimmermann, 2000). Theoretically, the DIG-1 predicted polypeptide could extend over 100 nm in length, if one takes into account the approximate size of each of its domains and assumes that domains are ordered like beads on a string. This size would be sufficient to bridge the extracellular matrix that separates neurons from adjacent muscle cells from which *dig-1* has been shown to be secreted through mosaic analysis. With cell adhesion and cell–cell interaction domains at either end of the protein, DIG-1 could straddle the basement membrane and interact with neurons on one end of the protein, and muscle and/or epidermal cell surfaces on the other end. DIG-1 may therefore be viewed as a scaffold that keeps individual neurons organized. As mentioned above, this putative scaffolding function is not required during development of neurons, but rather required to maintain structural organization.

3.3.3. EGL-15(5A): A specific isoform of the FGF receptor mediates axonal maintenance

egl-15 encodes the *C. elegans* fibroblast growth factor receptor tyrosine kinase and harbors three immunoglobulin domains in its extracellular domain. A specific splice form of the receptor, defined by the inclusion of an additional extracellular segment between the first and second Ig domain (Fig. 6.3), termed EGL-15(5A), is specifically required for maintenance of the position of axons, and has no role in axon guidance like other EGL-15 isoforms (Bülow *et al.*, 2004). In contrast to *dig-1* and *sax-7*, neuron soma position is unaffected and a more restricted number of axons are affected in the ventral nerve cord, compared to *sax-7* and *dig-1* mutants. This observation reveals an important feature of maintenance factors—they appear to be cell-type specific, with some neurons requiring all known maintenance factors and others only a subset (plus perhaps other as yet unknown factors) (Pocock *et al.*, 2008).

The maintenance function of EGL-15 is independent of its canonical signaling role, as neither its cognate FGF ligands nor its kinase domain are required for its maintenance role (Bülow *et al.*, 2004). Together with its focus of action in the epidermis, which underlies the ventral nerve cord, it is conceivable that EGL-15 may act in an adhesive function, possibly providing some adhesive substratum that keeps axons in place. It is likely that in such a role, EGL-15 is part of a larger adhesive complex, since, curiously, the extracellular domain of EGL-15 alone is sufficient to provide a maintenance role (Bülow *et al.*, 2004). An alternative scenario would be that EGL-15 is—like Eph-type receptor tyrosine kinases (Klein, 2009)—involved in some "inside–out," reverse signaling events with unknown receptors localized, for example, on axonal surfaces.

3.3.4. ZIG proteins

Eight genes in the *C. elegans* genome, named the *zig* genes, encode small proteins with two immunoglobulin domains (Fig. 6.3). All except one (ZIG-1) contain no transmembrane domain and are predicted to be secreted, likely from a unilateral neuron in the ventral nerve cord, called PVT (Aurelio *et al.*, 2002). Loss of individual *zig* genes or combinations of *zig* genes result in a diverse set of phenotypes. Loss of *zig-4* function leads to defects in axon maintenance in the ventral nerve cord, but has no effect on the position of neuronal soma (Aurelio *et al.*, 2002). The initial position adopted by the axons during the establishment of the ventral nerve cord in the developing embryo is normal, underscoring that the wild-type *zig-4* gene is dedicated to maintaining axons in their appropriate position. Consistent with its restricted role in maintenance, and in contrast to the molecules discussed above, *zig-4* appears to be expressed only postembryonically, supporting its apparently dedicated role as a maintenance factor. The spectrum of phenotypic defects of *zig-4* mutants is much more restricted than that of *dig-1* and *sax-7* and overlapping but still distinct from that of *egl-15* mutants, underscoring the cell-type specificity of axon maintenance mechanisms.

Other *zig* genes also appear to function in maintenance of neuronal architecture. For example, *zig-3* displays similar defects in ventral cord axon positioning as *zig-4* mutants, suggesting that these two molecules may act together (C. Bénard and O. Hobert, unpublished results). In contrast, a double knockout of the *zig-5* and *zig-8* genes results in neuronal soma position defects in head ganglia, mimicking those observed in *sax-7* and *dig-1* mutants (C. Bénard and O. Hobert, unpublished results).

Numerous proteins characterized by two-Ig domains exist in other metazoans such as flies and mammals, which could possibly be functional homologues. In flies, there are at least 20 genes encoding two-Ig domain proteins (Rougon and Hobert, 2003). The secreted Beaten path or Beat Ia, and a subset of 13 transmembrane Beat-like proteins, function to regulate fasciculation during the development of the *Drosophila* nervous system (Fambrough and Goodman, 1996; Pipes *et al.*, 2001). Possible maintenance functions of the Beat proteins have not yet been explored. Another group of ~20 small two-Ig proteins, some with, some without a transmembrane domain, is the Dpr-Ig family, defined by the protein Dpr1 (defective in proboscis extension response) (Nakamura *et al.*, 2002). Dpr1 is required for proper behavioral response to salt (Nakamura *et al.*, 2002; Woo *et al.*, 2008).

Whether these defects are due to maintenance defects has not been investigated yet. Mammalian genomes contain a multitude of genes coding for two-Ig domain-containing proteins, whose function in nervous system maintenance is also unexplored.

3.3.5. F-spondin/SPON-1: An extracellular matrix protein

Spon-1, the worm homolog of the vertebrate F-spondin protein, is a component of the basement membrane of the worm's body wall muscles and pharynx. *spon-1* is required for axons of ventral cord neurons to maintain their position (Woo *et al.*, 2008). In contrast to the factors mentioned above, *spon-1* also has functions during axonal development. In *spon-1* mutants, developing axons are misguided and there are major fasciculation defects, making it more difficult to separate the worsening of developmental defects from subsequent maintenance failure. *spon-1* also display complex genetic interactions with *zig-4* and *egl-15*, which are not presently understood, but further corroborate the importance of the extracellular matrix in maintaining neuronal structures.

4. Discussion

Genetic analysis in *C. elegans* has revealed a surprisingly complex machinery that serves to maintain nervous system architecture. The normal function of maintenance factors is to counteract specific mechanical challenges exerted onto a nervous system. As such, their importance only becomes apparent if the entire nervous system is looked at as a whole, on the system level, rather than at the isolated cell level. Notably, though, while movement is *required* to induce defects in the absence of maintenance factors, it is not in all cases *sufficient* as maintenance defects become apparent only at stages where mechanical stress coincides with remodeling of the nervous system. That is, even though mutant embryos that lack maintenance factors already move vigorously right before hatching, most axonal and soma maintenance defects become apparent only by the first larval stage, when new neurons and their axons are added to the ventral nerve cord and when the underlying epidermis gets remodeled through cell fusions. It is therefore a combination of various external, destabilizing influences (mechanical stress, remodeling, growth) that necessitate the existence of maintenance factors. Thus, maintenance factors may not only be viewed as cell-specific, mechanical glues, but also as molecules that shield pre-existing neuronal structures from potentially disruptive influences of guidance cues that are required to integrate newly generated neurons into pre-existing circuitries.

Maintenance factors affect a distinct, though largely overlapping spectrum of axons and neuronal cell bodies, and are supplied by distinct cell types.

ZIG-4 is secreted from the PVT interneuron to affect the maintenance of several sets of axons (Aurelio *et al.*, 2002); EGL-15 acts in the epidermis on which the axons are positioned (Bülow *et al.*, 2004); DIG-1 is secreted from muscle (Benard *et al.*, 2006); and SAX-7 acts within neurons to affect axon position (Pocock *et al.*, 2008; Sasakura *et al.*, 2005). These neuronal maintenance factors, all bearing Ig domains, as well as possibly a number of unidentified factors, likely interact in different, cell-type-specific combinations to build adhesive complexes that anchor neurons and their projections in their appropriate environment. They might directly mediate, or regulate through signaling events, direct cell–cell interactions among neurons, between neurons and other neighboring cells, or with the extracellular matrix. DIG-1 may be a central component to all distinct maintenance mechanisms as its loss affects the most types of neurons. In light of its size and large number of interaction domains, DIG-1 may nucleate the assembly of cell-specific maintenance complexes. Secreted proteins such as ZIG-4 may contribute in building or facilitating the building of such complexes.

Loss of activity of different maintenance factors leads to neuron-specific defects, at different time points during the animal's life, reflecting the precise requirements of different cells in particular temporal and tissue contexts. The time of occurrence and degree of displacement of a neuron might depend on the age of a neuron and on the strength of adhesion or attachment to its environment, as well as on the developmental or mechanical events affecting the neuron, and the specific tissue topology. Two broad categories of temporal requirements for architectural maintenance factors can be distinguished, one relating to specific stresses during critical time windows, and the other relating to general long-term stability throughout life. The first temporal requirement is that of a critical period or time window in development or maturation of the nervous system, where a neuron becomes displaced in the absence of a maintenance factor, losing the proper position it had initially acquired. The state of affairs remains fixed after this time window passes. This is best illustrated by the axon that flips-over in the ventral nerve cord due to body movements during a time when instability is generated in the vicinity of the axons by the addition of motor neurons and the remodeling of the epidermis. After this time, while the strain from body movements of locomotion persists, the remodeled epidermis constitutes a physical barrier that prevents further axon displacement (Fig. 6.2D). The second temporal requirement for architectural maintenance is manifested throughout life and ensures long-term architecture stability. A lack of this continuous maintenance action relates to the age-dependent progressive accumulation of defects. This is well illustrated by the displaced chemosensory neurons in the head ganglia in *dig-1* and *sax-7* mutants, which show the first signs of displacement in the second larval stage and worsen over time with the continued movements of the animal. Nervous system structures that show later onsets of progressive maintenance defects might be better anchored in their surrounding environment, therefore taking more time to be displaced by mechanical stress.

It is important to emphasize that the discovery of dedicated maintenance factors is contrary to a perhaps more parsimonious model in which one may have assumed that factors that help develop a system merely "stick around" to ensure that neuronal ensembles, once established, maintain their correct architecture. The existence of factors that appear to be entirely dedicated to a maintenance role obliterates such a simplistic model. The need for dedicated maintenance mechanisms, rather than a reutilization and/or continuous use of developmental cues, could be explained by a possibly disruptive impact that continuously expressed guidance cues may have on newly developing neurons. Indeed, molecules involved in axon guidance are often expressed transiently during development. It is, in fact, the transient, dynamic, and localized expression of a limited set of guidance cues that is thought to generate complex patterns of neuronal connectivities. Continuous expression to maintain circuitries may therefore be undesired. Whether this is the rule or the exception remains to be determined, as several molecules known to maintain cell-intrinsic features of a neuron or appropriate contacts with neighbors, such as synaptic adhesion molecules (Sanes and Lichtman, 1999), have roles in development as well.

A distinction between molecular mechanisms that "initiate" and "maintain" a biological process is a concept that is not unique to the regulation of structural aspects of the nervous system. In the field of gene regulation, it is also known that given genetic loci require distinct factors (DNA-binding transcription factors) for the initiation of a specific expression pattern and a distinct set (chromatin-modifying factors and other transcription factors) for ensuing maintenance (Brock and Fisher, 2005). Such separations may abound in other biological processes as well.

An understanding of maintenance mechanisms holds the prospect of uncovering new clues about neurodegenerative disease. It is conceivable that some neurodegenerative diseases, for which the causes are unknown, may be the result of postdevelopmental failures in the maintenance mechanisms. Conceivably, the displacement of neurons and axons results in subsequent demise of neurons, loss of their function, eventually bringing about neurological symptoms. Human homologs of some of the genes identified in *C. elegans* may be candidates for the diagnosis and treatment of some neurodegenerative disorders.

ACKNOWLEDGMENTS

We thank the members of the Hobert lab and Carol Mason for discussion on maintenance and comments on the manuscript, as well as Iva Greenwald, Chris Henderson, Wes Grueber, and Brian McCabe for inspiring discussions. C.B. was funded by postdoctoral funds from the

Natural Sciences and Engineering Research Council of Canada and the Canadian Institute of Health Research. This work was funded, in part, from a grant by the Muscle Dystrophy Association. O.H. is an Investigator of the HHMI.

REFERENCES

Aurelio, O., Hall, D. H., and Hobert, O. (2002). Immunoglobulin-domain proteins required for maintenance of ventral nerve cord organization. *Science* **295,** 686–690.

Axang, C., Rauthan, M., Hall, D. H., and Pilon, M. (2007). The twisted pharynx phenotype in *C. elegans*. *BMC Dev. Biol.* **7,** 61.

Bandtlow, C. E., and Zimmermann, D. R. (2000). Proteoglycans in the developing brain: New conceptual insights for old proteins. *Physiol. Rev.* **80,** 1267–1290.

Benard, C. Y., Boyanov, A., Hall, D. H., and Hobert, O. (2006). DIG-1, a novel giant protein, non-autonomously mediates maintenance of nervous system architecture. *Development* **133,** 3329–3340.

Boulin, T., Pocock, R., and Hobert, O. (2006). A novel Eph receptor-interacting IgSF protein provides *C. elegans* motoneurons with midline guidepost function. *Curr. Biol.* **16,** 1871–1883.

Brock, H. W., and Fisher, C. L. (2005). Maintenance of gene expression patterns. *Dev. Dyn.* **232,** 633–655.

Brown, M. C., Holland, R. L., and Hopkins, W. G. (1981). Motor nerve sprouting. *Ann. Rev. Neurosci.* **4,** 17–42.

Bülow, H. E., Boulin, T., and Hobert, O. (2004). Differential functions of the *C. elegans* FGF receptor in axon outgrowth and maintenance of axon position. *Neuron* **42,** 367–374.

Burket, C. T., Higgins, C. E., Hull, L. C., Berninsone, P. M., and Ryder, E. F. (2006). The *C. elegans* gene dig-1 encodes a giant member of the immunoglobulin superfamily that promotes fasciculation of neuronal processes. *Dev. Biol.* **299,** 193–205.

Cao, L., Dhilla, A., Mukai, J., Blazeski, R., Lodovichi, C., Mason, C. A., and Gogos, J. A. (2007). Genetic modulation of BDNF signaling affects the outcome of axonal competition *in vivo*. *Curr. Biol.* **17,** 911–921.

Chen, L., Ong, B., and Bennett, V. (2001). LAD-1, the *Caenorhabditis elegans* L1CAM homologue, participates in embryonic and gonadal morphogenesis and is a substrate for fibroblast growth factor receptor pathway-dependent phosphotyrosine-based signaling. *J. Cell. Biol.* **154,** 841–856.

Coburn, C. M., Mori, I., Ohshima, Y., and Bargmann, C. I. (1998). A cyclic nucleotide-gated channel inhibits sensory axon outgrowth in larval and adult *Caenorhabditis elegans*: A distinct pathway for maintenance of sensory axon structure. *Development* **125,** 249–258.

Cook, B., Hardy, R. W., McConnaughey, W. B., and Zuker, C. S. (2008). Preserving cell shape under environmental stress. *Nature* **452,** 361–364.

Emoto, K., Parrish, J. Z., Jan, L. Y., and Jan, Y. N. (2006). The tumour suppressor Hippo acts with the NDR kinases in dendritic tiling and maintenance. *Nature* **443,** 210–213.

Fambrough, D., and Goodman, C. S. (1996). The *Drosophila* beaten path gene encodes a novel secreted protein that regulates defasciculation at motor axon choice points. *Cell* **87,** 1049–1058.

Florence, S. L., Taub, H. B., and Kaas, J. H. (1998). Large-scale sprouting of cortical connections after peripheral injury in adult macaque monkeys. *Science* **282,** 1117–1121.

Fox, M. A., Sanes, J. R., Borza, D. B., Eswarakumar, V. P., Fassler, R., Hudson, B. G., John, S. W., Ninomiya, Y., Pedchenko, V., Pfaff, S. L., Rheault, M. N., Sado, Y., *et al.* (2007). Distinct target-derived signals organize formation, maturation, and maintenance of motor nerve terminals. *Cell* **129,** 179–193.

Fransen, E., Van Camp, G., Vits, L., and Willems, P. J. (1997). L1-associated diseases: Clinical geneticists divide, molecular geneticists unite. *Hum. Mol. Genet.* **6,** 1625–1632.

Fruttiger, M., Montag, D., Schachner, M., and Martini, R. (1995). Crucial role for the myelin-associated glycoprotein in the maintenance of axon-myelin integrity. *Eur. J. Neurosci.* **7,** 511–515.

Gorski, J. A., Zeiler, S. R., Tamowski, S., and Jones, K. R. (2003). Brain-derived neurotrophic factor is required for the maintenance of cortical dendrites. *J. Neurosci.* **23,** 6856–6865.

Griffiths, I., Klugmann, M., Anderson, T., Yool, D., Thomson, C., Schwab, M. H., Schneider, A., Zimmermann, F., McCulloch, M., Nadon, N., and Nave, K. A. (1998). Axonal swellings and degeneration in mice lacking the major proteolipid of myelin. *Science* **280,** 1610–1613.

Henderson, C. E. (1996). Role of neurotrophic factors in neuronal development. *Curr. Opin. Neurobiol.* **6,** 64–70.

Hubel, D. H., and Wiesel, T. N. (1970). The period of susceptibility to the physiological effects of unilateral eye closure in kittens. *J. Physiol.* **206,** 419–436.

Huberman, A. D., Feller, M. B., and Chapman, B. (2008). Mechanisms underlying development of visual maps and receptive fields. *Annu. Rev. Neurosci.* **31,** 479–509.

Jessen, K. R. (2004). Glial cells. *Int. J. Biochem. Cell. Biol.* **36,** 1861–1867.

Kaufmann, W. E., and Moser, H. W. (2000). Dendritic anomalies in disorders associated with mental retardation. *Cereb. Cortex* **10,** 981–991.

Klein, R. (2009). Bidirectional modulation of synaptic functions by Eph/ephrin signaling. *Nat. Neurosci.* **12,** 15–20.

Klugmann, M., Schwab, M. H., Puhlhofer, A., Schneider, A., Zimmermann, F., Griffiths, I. R., and Nave, K. A. (1997). Assembly of CNS myelin in the absence of proteolipid protein. *Neuron* **18,** 59–70.

Knobel, K. M., Davis, W. S., Jorgensen, E. M., and Bastiani, M. J. (2001). UNC-119 suppresses axon branching in *C. elegans*. *Development* **128,** 4079–4092.

LaMonte, B. H., Wallace, K. E., Holloway, B. A., Shelly, S. S., Ascano, J., Tokito, M., Van Winkle, T., Howland, D. S., and Holzbaur, E. L. (2002). Disruption of dynein/dynactin inhibits axonal transport in motor neurons causing late-onset progressive degeneration. *Neuron* **34,** 715–727.

Law, J. W., Lee, A. Y., Sun, M., Nikonenko, A. G., Chung, S. K., Dityatev, A., Schachner, M., and Morellini, F. (2003). Decreased anxiety, altered place learning, and increased CA1 basal excitatory synaptic transmission in mice with conditional ablation of the neural cell adhesion molecule L1. *J. Neurosci.* **23,** 10419–10432.

Lenroot, R. K., and Giedd, J. N. (2006). Brain development in children and adolescents: Insights from anatomical magnetic resonance imaging. *Neurosci. Biobehav. Rev.* **30,** 718–729.

Lindahl, U., and Hook, M. (1978). Glycosaminoglycans and their binding to biological macromolecules. *Annu. Rev. Biochem.* **47,** 385–417.

Loria, P. M., Hodgkin, J., and Hobert, O. (2004). A conserved postsynaptic transmembrane protein affecting neuromuscular signaling in Caenorhabditis elegans. *J. Neurosci.* **24,** 2191–2201.

Marrs, G. S., Honda, T., Fuller, L., Thangavel, R., Balsamo, J., Lilien, J., Dailey, M. E., and Arregui, C. (2006). Dendritic arbors of developing retinal ganglion cells are stabilized by beta 1-integrins. *Mol. Cell. Neurosci.* **32,** 230–241.

Moresco, E. M., Donaldson, S., Williamson, A., and Koleske, A. J. (2005). Integrin-mediated dendrite branch maintenance requires Abelson (Abl) family kinases. *J. Neurosci.* **25,** 6105–6118.

Nakamura, M., Baldwin, D., Hannaford, S., Palka, J., and Montell, C. (2002). Defective proboscis extension response (DPR), a member of the Ig superfamily required for the gustatory response to salt. *J. Neurosci.* **22,** 3463–3472.

Peckol, E. L., Zallen, J. A., Yarrow, J. C., and Bargmann, C. I. (1999). Sensory activity affects sensory axon development in *C. elegans*. *Development* **126,** 1891–1902.

Peng, Y. W., Zallocchi, M., Meehan, D. T., Delimont, D., Chang, B., Hawes, N., Wang, W., and Cosgrove, D. (2008). Progressive morphological and functional defects in retinas from alpha1 integrin-null mice. *Invest. Ophthalmol. Vis. Sci.* **49,** 4647–4654.

Pipes, G. C., Lin, Q., Riley, S. E., and Goodman, C. S. (2001). The Beat generation: A multigene family encoding IgSF proteins related to the Beat axon guidance molecule in Drosophila. *Development* **128,** 4545–4552.

Pocock, R., Benard, C. Y., Shapiro, L., and Hobert, O. (2008). Functional dissection of the *C. elegans* cell adhesion molecule SAX-7, a homologue of human L1. *Mol. Cell. Neurosci.* **37,** 56–68.

Rougon, G., and Hobert, O. (2003). New Insights into the Diversity and Function of Neuronal Immunoglobulin Superfamily Molecules. *Annu. Rev. Neurosci.* **26,** 207–238.

Sanes, J. R., and Lichtman, J. W. (1999). Development of the vertebrate neuromuscular junction. *Annu. Rev. Neurosci.* **22,** 389–442.

Sanes, J. R., and Lichtman, J. W. (2001). Induction, assembly, maturation and maintenance of a postsynaptic apparatus. *Nat. Rev. Neurosci.* **2,** 791–805.

Sasakura, H., Inada, H., Kuhara, A., Fusaoka, E., Takemoto, D., Takeuchi, K., and Mori, I. (2005). Maintenance of neuronal positions in organized ganglia by SAX-7, a Caenorhabditis elegans homologue of L1. *Embo J.* **24,** 1477–1488.

Shirasaki, R., and Pfaff, S. L. (2002). Transcriptional codes and the control of neuronal identity. *Annu. Rev. Neurosci.* **25,** 251–281.

Tailby, C., Wright, L. L., Metha, A. B., and Calford, M. B. (2005). Activity-dependent maintenance and growth of dendrites in adult cortex. *Proc. Natl. Acad. Sci. USA* **102,** 4631–4636.

Toga, A. W., Thompson, P. M., and Sowell, E. R. (2006). Mapping brain maturation. *Trends Neurosci.* **29,** 148–159.

Trachtenberg, J. T., Chen, B. E., Knott, G. W., Feng, G., Sanes, J. R., Welker, E., and Svoboda, K. (2002). Long-term *in vivo* imaging of experience-dependent synaptic plasticity in adult cortex. *Nature* **420,** 788–794.

Van Essen, D. C. (1997). A tension-based theory of morphogenesis and compact wiring in the central nervous system. *Nature* **385,** 313–318.

Wang, X., Kweon, J., Larson, S., and Chen, L. (2005). A role for the *C. elegans* L1CAM homologue lad-1/sax-7 in maintaining tissue attachment. *Dev. Biol.* **284,** 273–291.

White, J. G., Southgate, E., Thomson, J. N., and Brenner, S. (1986). The structure of the nervous system of the nematode *Caenorhabditis elegans. Philosophical Transactions of the Royal Society of London B. Biological Sciences* **314,** 1–340.

Woo, W. M., Berry, E. C., Hudson, M. L., Swale, R. E., Goncharov, A., and Chisholm, A. D. (2008). The *C. elegans* F-spondin family protein SPON-1 maintains cell adhesion in neural and non-neural tissues. *Development* **135,** 2747–2756.

Xu, B., Zang, K., Ruff, N. L., Zhang, Y. A., McConnell, S. K., Stryker, M. P., and Reichardt, L. F. (2000). Cortical degeneration in the absence of neurotrophin signaling: Dendritic retraction and neuronal loss after removal of the receptor TrkB. *Neuron* **26,** 233–245.

Yin, D. M., Huang, Y. H., Zhu, Y. B., and Wang, Y. (2008). Both the establishment and maintenance of neuronal polarity require the activity of protein kinase D in the Golgi apparatus. *J. Neurosci.* **28,** 8832–8843.

Zallen, J. A., Kirch, S. A., and Bargmann, C. I. (1999). Genes required for axon pathfinding and extension in the *C. elegans* nerve ring. *Development* **126,** 3679–3692.

Zhao, H., and Nonet, M. L. (2000). A retrograde signal is involved in activity-dependent remodeling at a *C. elegans* neuromuscular junction. *Development* **127,** 1253–1266.

Zhou, S., Opperman, K., Wang, X., and Chen, L. (2008). unc-44 Ankyrin and stn-2 gamma-syntrophin regulate sax-7 L1CAM function in maintaining neuronal positioning in *Caenorhabditis elegans. Genetics* **180,** 1429–1443.

Subject Index

T

U

V

W

X

Z

Contents of Previous Volumes

Volume 47

Volume 48

Volume 49

Volume 50

Volume 51

Volume 52

Volume 53

Volume 54

Volume 55

Volume 56

Volume 57

Volume 60

Volume 61

Volume 62

Volume 63

Volume 64

Volume 65

Volume 66

Volume 67

Volume 68

Volume 69

Volume 70

Volume 73

Volume 74

Volume 75

Volume 76

Volume 77

Volume 80

Volume 81

Volume 82

Volume 83

Volume 84

Volume 85

Volume 86

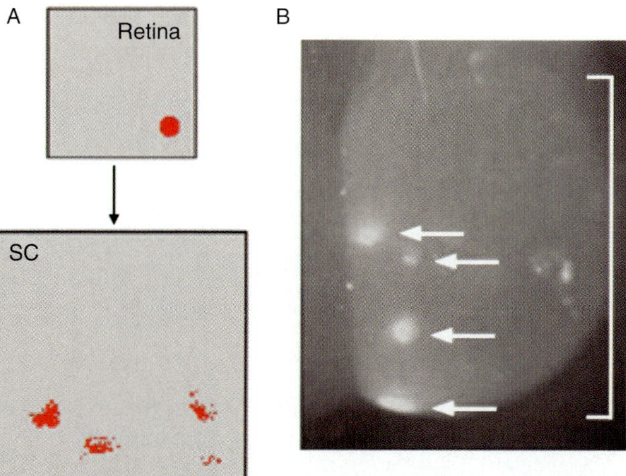

Hugh Simpson *et al.*, Figure 1.11 Simulations and experimental results in ephrin misexpression studies. (A) Simulated injection of tracer dye projects to the superior colliculus (SC) in the probabilistic sorting model (Tsigankov and Koulakov, 2006) applied to an ephrin misexpression study. (B) Corresponding experimental results for the ephrin misexpression study (Feldheim *et al.*, 2000) where focal retinal injections of DiI into the retina (similarly to the simulations shown in (A) but the experimental retinal injection is not shown) resulted in multiple termination zones (arrows) in knock-out animals ((A) from Tsigankov and Koulakov, 2006; adapted with permission from Springer Science + Business Media. (B) from Feldheim *et al.*, 2000; reproduced with permission from Elsevier)

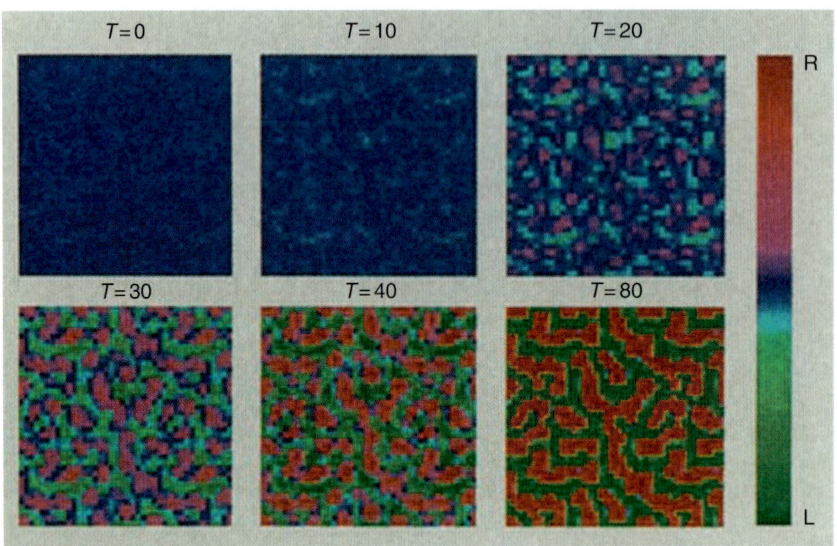

Hugh Simpson *et al.*, Figure 1.15 Time development of ocular dominance columns in the model of Miller *et al.* (1989). Segregation from an initially unsegregated state can be seen as development proceeds. Developmental time is measured in arbitrary units (from Miller *et al.*, 1989; reproduced with permission from AAAS).

Hugh Simpson *et al.*, Figure 1.18 Elastic net simulations of visual map development. Left panel: Ocular dominance map. Middle panel: Orientation preference map. Right panel: Superimposed contours of ocular dominance and orientation. Note the tendency to intersect at steep angles, and for pinwheels to lie at the center of ocular dominance columns (see Carreira-Perpiñán *et al.*, 2005 for more details).

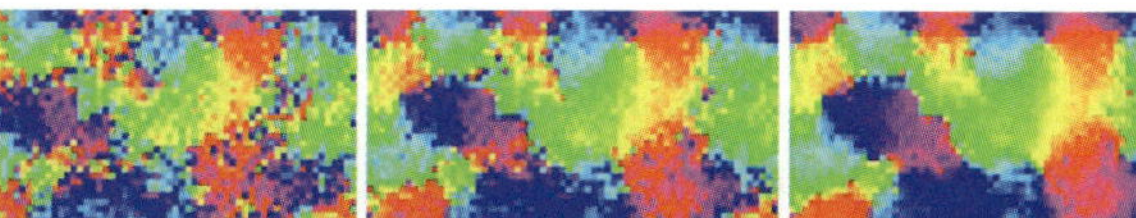

Hugh Simpson *et al.*, Figure 1.19 Development of a direction preference map in the STDP-based model of Wenisch *et al.* (2005). Sequence of map development at four different time points. Colors corresponding to different direction preferences are shown above the simulations (from Wenisch *et al.*, 2005; reproduced with permission from Springer Science + Business Media).

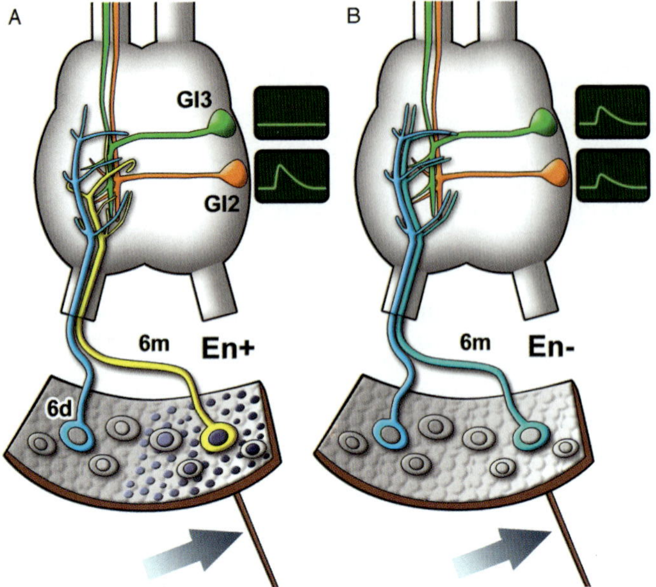

Daniel A. Colón-Ramos, Figure 2.1 Transcription factor *Engrailed* directs synaptic specificity in the cockroach cercal system. Wiring diagram of the cockroach cercal system in the second instar cockroach. (A) In wild-type animals, medial sensory neuron (neuron 6m, in yellow) expresses Engrailed (represented by dark nuclei), while lateral sensory neuron (6d, in blue) does not (represented by clear nuclei). Although the axonal arbors of the medial and lateral sensory neurons overlap, they display specificity by connecting with specific target interneurons: medial sensory neuron connects to interneuron Gl2 (in orange), while lateral sensory neuron connects to interneuron Gl3 (in green). This specificity can be physiologically recorded, so that, for instance, stimuli in the filiform hair linked to the medial sensory neuron (dark arrow) results in depolarization of interneuron Gl2 and not Gl3 (represented by physiological recording to the right of the schematic). (B) Loss of Engrailed by dsRNA disrupts the medial sensory neuron identity and connectivity. In the absence of Engrailed, the medial sensory neuron (6m) adopts an identity similar to the lateral sensory neuron in terms of the branching structure (compare branching schematic of 6m in A (yellow) with 6m in B (blue)). Loss of Engrailed also disrupts medial sensory neuron synaptic specificity (represented by physiological recording to the right of the schematic).

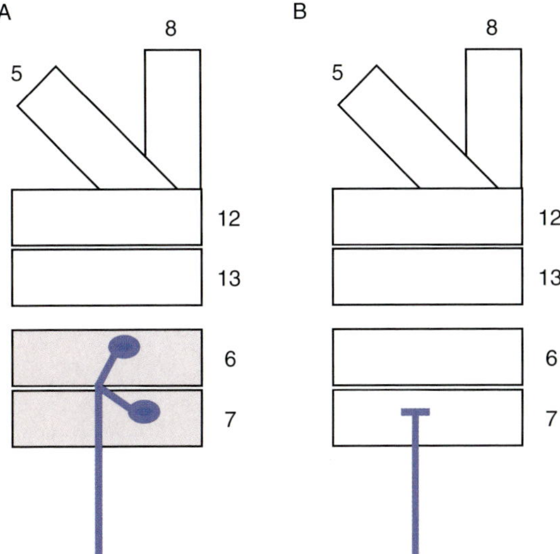

Daniel A. Colón-Ramos, Figure 2.2 Axon guidance molecule Netrin is required for synaptic targeting events in the *Drosophila* embryo. Schematic diagram of RP3 neuron (blue) and the body wall muscles in the *Drosophila* embryo (represented here are muscles 6, 7, 13, 12, 5, and 8). (A) In wild-type animals, muscles 6 and 7 express Netrin (in pink). Muscles 12, 13, 5, and 8 express repulsive cues such as semaphorins, and do not express Netrin (lack of Netrin expression represented in white). Expression of Netrin by muscles 6 and 7 induces short-range targeting and specific innervation of these muscles by the RP3 neuron. (B) In Netrin loss-of-function mutants, the RP3 neuron reaches the neighborhood of muscles 6 and 7 in a timely fashion, coming within filopodial reach of its targets, but fails to innervate these muscles correctly. These studies suggest that Netrin is not required for long-range guidance of the RP3 neuron, but is instead required for its short-range synaptic targeting.

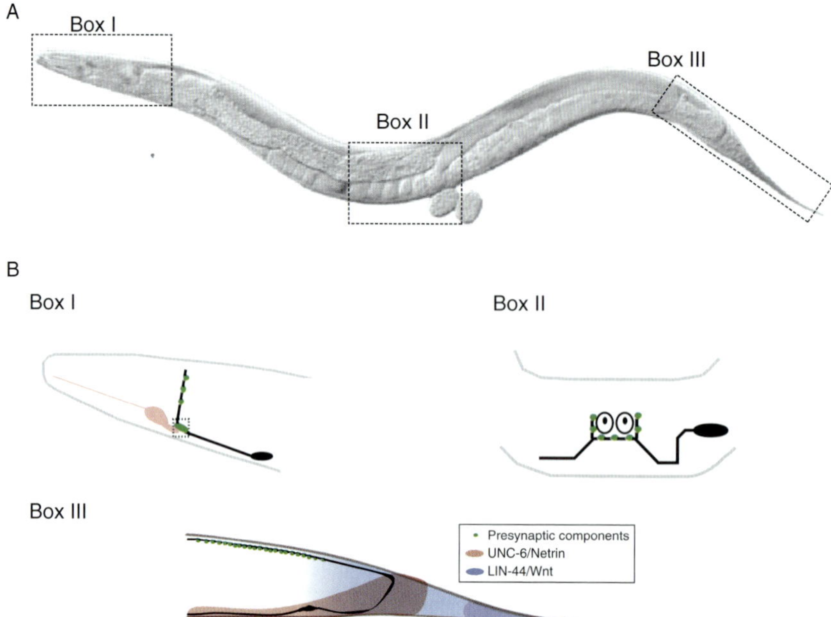

Daniel A. Colón-Ramos, Figure 2.3 Guidepost cells direct synaptic specification. Schematic representing examples of guidepost cells directing synaptic specification in *C. elegans*. (A) Image of *C. elegans* with the regions discussed boxed. (B) *Box-I*: In the nematode head, interneuron AIY (black) contacts many neurons, but connects specifically to interneuron RIA (not shown) at a subcellular region of its neurite (boxed). This specificity is directed by ventral cephalic sheath cells (pink). Ventral cephalic sheath cells are glial cells that project a process posteriorly, where it contacts AIY and RIA, and also express Netrin. The expression of Netrin directs presynaptic assembly (in green) in the correct subcellular region of AIY (boxed region). *Box-II*: In the nematode egg-laying circuit, neuron HSNL (in black) innervates other neurons and muscles (not shown) in a specific and stereotyped fashion. The postsynaptic partners of HSNL are not required for the precise assembly of presynaptic specializations (in green) in a subcellular region of HSNL. Instead, guidepost epithelial cells (dark spheres) express an immunoglobulin superfamily receptor (SYG-2) that directs where synapses form in HSNL. *Box-III*: In the posterior part of the nematode, neuron DA9 (in black) elaborates a dendrite anteriorly within the ventral nerve cord and extends an axon commissurally and then longitudinally along the dorsal nerve cord. UNC-6/Netrin (pink) and LIN-44/Wnt (blue) direct synaptic specification (green) by inhibiting formation of synapses from discrete subcellular domains. Expression of *lin-44/wnt* (blue) by cells in the posterior part of the animal prevents ectopic synapse formation in the commissure, while expression of Netrin (pink) by ventral cells excludes presynaptic components from the ventral dendrite.

Daniel A. Colón-Ramos, Figure 2.4 Bergmann glia direct the innervation of Stellate axons to the Purkinje dendrites. Purkinje neurons (yellow) are innervated by stellate interneurons (green) exclusively at the dendrites. This precision is directed by Bergmann glia (red), which act as guideposts by directing the stellate interneuron process to their Purkinje neuron targets (synapses in blue).

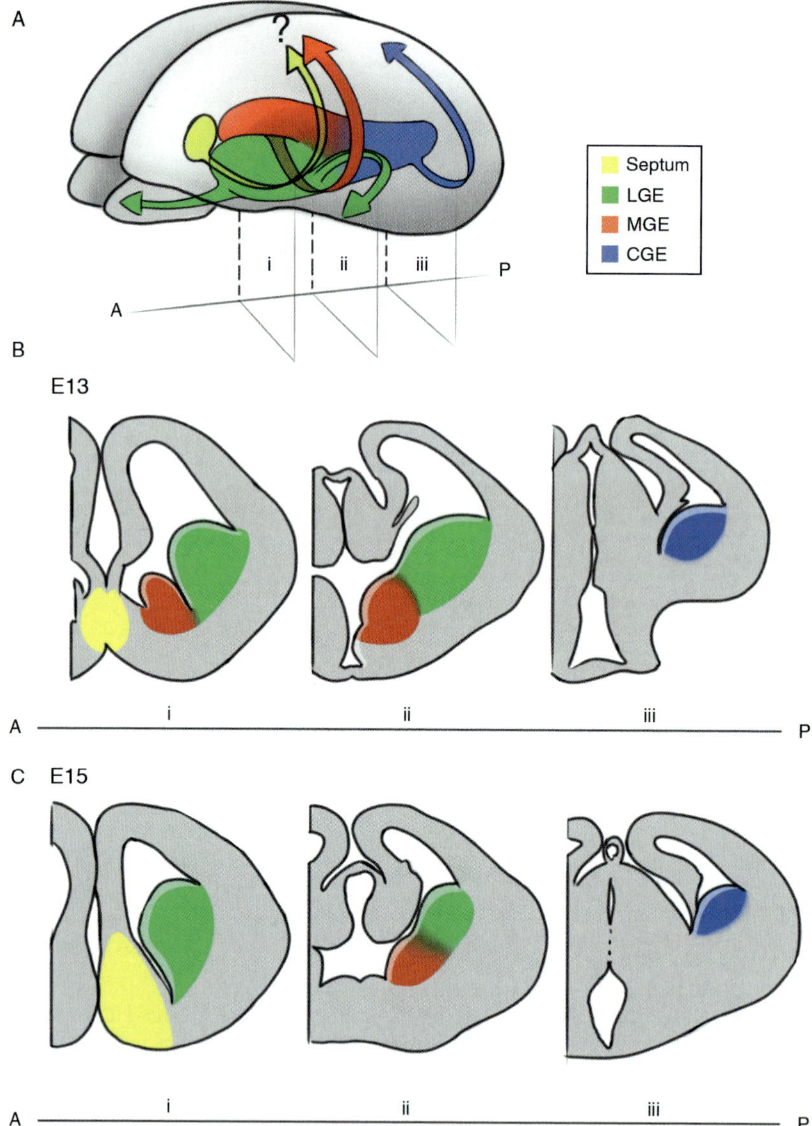

Renata Batista-Brito and Gord Fishell, Figure 3.1 Schematic representation of the developing murine brain. (A) Three-dimensional view of a developing brain indicating the ventral anatomical regions that give rise to interneurons and their correspondent migratory pathways. The LGE (green) generates interneurons that migrate to the olfactory bulb and striatum. Both the MGE (red) and CGE (blue) produce cortical interneurons. Whether the septum produces cortical interneurons (yellow) is still a matter of debate. (B–C) Coronal views of the brain in (A) at three locations (i–iii) along the anterior (A)–posterior (P) axis at the embryonic ages E13 (B) and E15 (C).

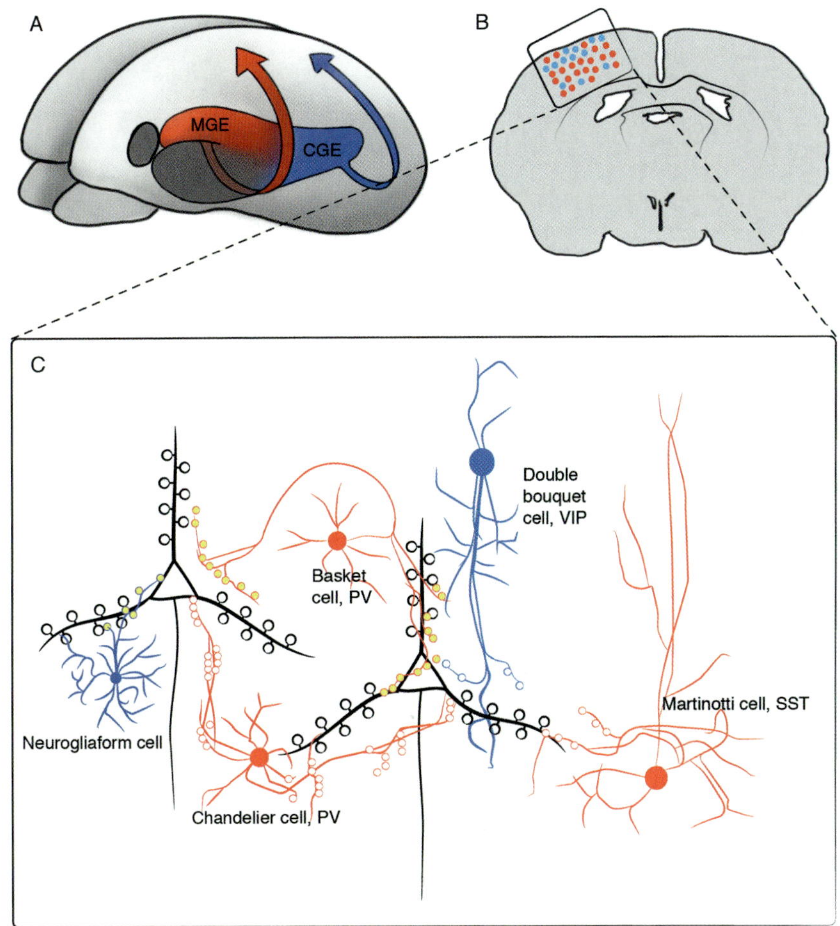

Renata Batista-Brito and Gord Fishell, Figure 3.2 Differential origin of cortical interneuron subtypes. (A) Three-dimensional view of an embryonic murine brain highlighting the two ventral regions that produce cortical interneurons, the MGE (red) and the CGE (blue). (B) Coronal view of an adult brain, illustrating the proportion and relative distribution of cortical interneurons derived from the MGE (red dots) and the CGE (blue dots). (C) Detailed schematic view of the boxed region in (B) illustrating the main interneuron subtypes derived from the MGE (red cells) and CGE (blue cells), and how they characteristically interact with the pyramidal cells (black cells). This figure is adapted from Kawaguchi and Kubota (2002).

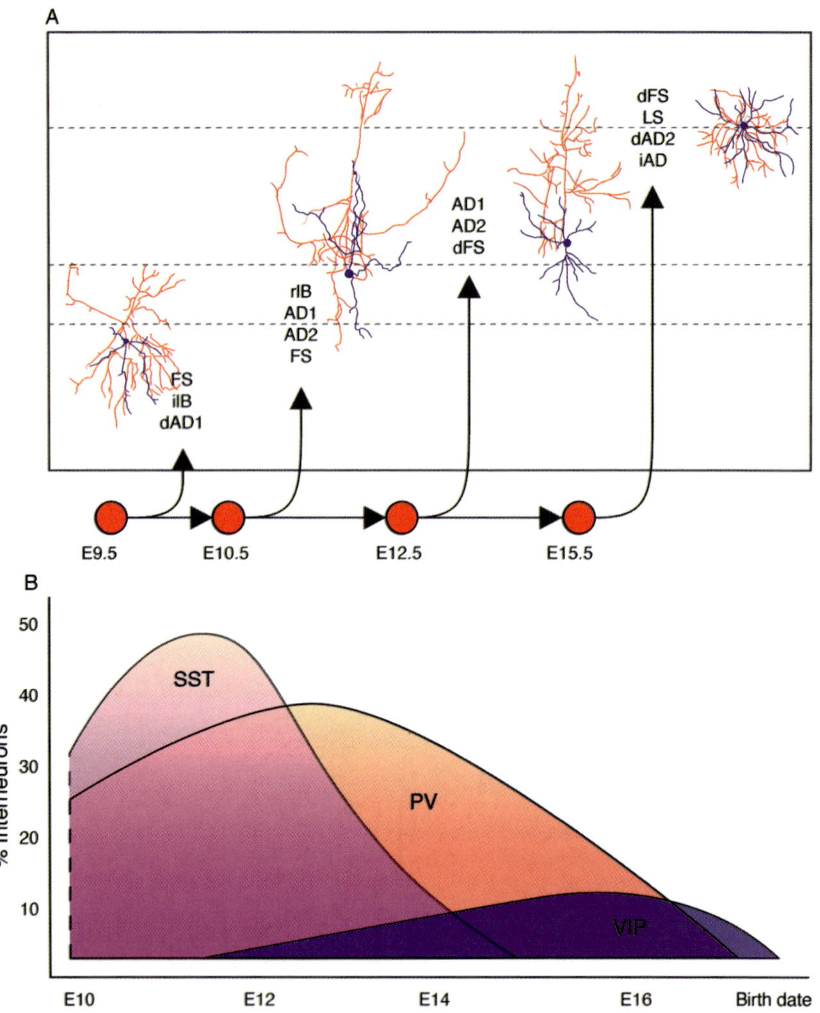

Renata Batista-Brito and Gord Fishell, Figure 3.3 Temporal origin of cortical interneurons. (A) Examples of the morphological and electrophysiological diversity of cortical interneurons derived from the MGE at different times. Axons and dendrites are indicated in red and blue, respectively. This figure is adapted from Miyoshi *et al.* (2007). (B) Diagram of temporal origin of three subtypes of MGE interneurons, the somatostatin (SST—pink), the parvalbumin (PV—orange), and a subtype of CGE-derived interneuron (VIP—blue).

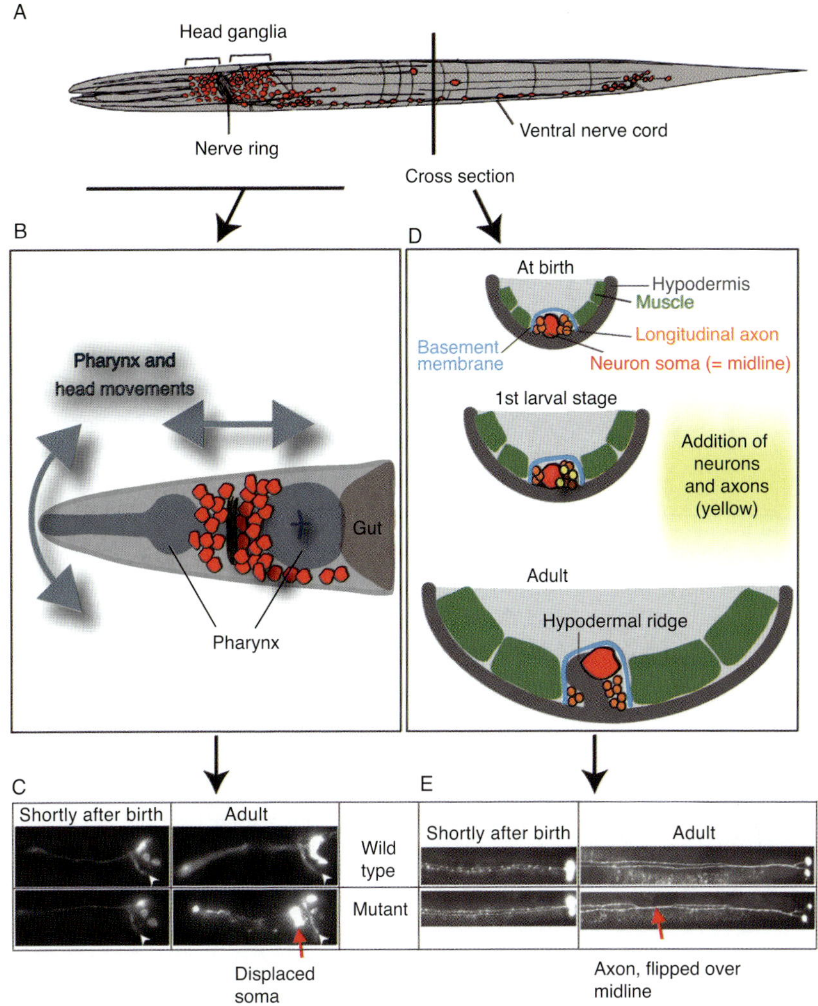

Claire Bénard and Oliver Hobert, Figure 6.2 Maintenance of neuronal architecture in the nervous system of *C. elegans*. (A) The majority of the 302 neurons (in red) of the worm are organized into major ganglia in the head and tail, as well as into fascicles such as the ventral nerve cord. This organization is established during embryogenesis and retained throughout life. (B) Schematic representation of the head of a worm, with neurons in head ganglia (in red). Throughout larval and adult stages, the worm swims in its environment foraging for food and feeds through the motions of its pharynx, which pumps food toward the gut. These lifelong motions exert considerable mechanical strain on the neurons situated near the pharynx and lead to the progressive displacement of neurons in the absence of dedicated neuronal maintenance mechanisms (C). (C) Example of maintenance defects in head ganglia. In wild-type worms, the cell bodies of a subset of chemosensory neurons of the head are located posterior to the nerve ring (indicated by a white arrowhead). This chemosensory architecture develops in embryogenesis and persists throughout life. In maintenance mutants such as *sax-7* or *dig-1*, the chemosensory neurons develop completely normally and are indistinguishable form wild-type at birth, but later become progressively displaced (red arrow, anterior to the nerve ring). (D) Diagram of the ventral portion of a cross section through a worm at birth,

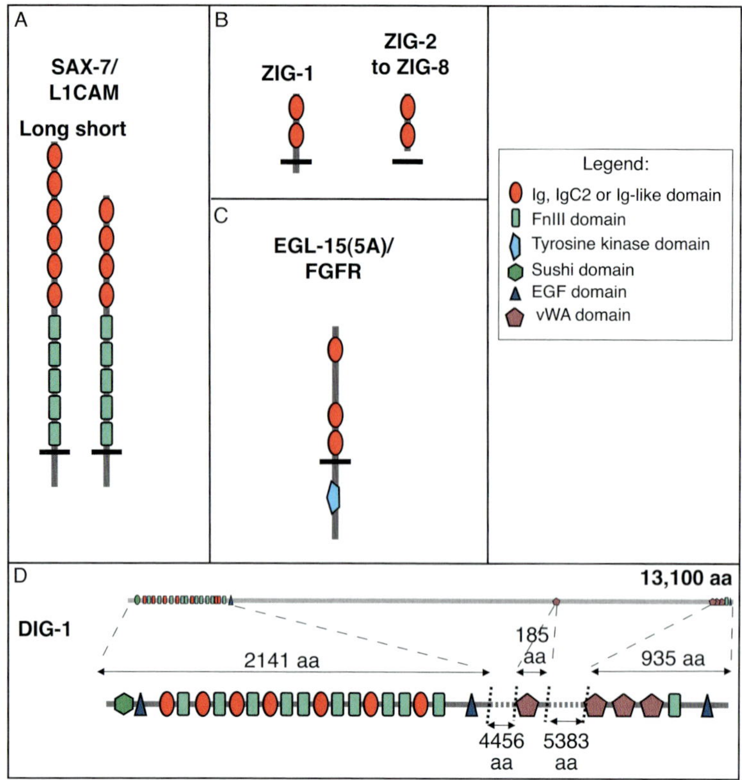

Claire Bénard and Oliver Hobert, Figure 6.3 Molecules implicated in maintenance of neuronal architecture. To date, diverse members of the immunoglobulin superfamily are implicated in maintaining the precise position of axons within fascicles and neuronal soma in ganglia in *C. elegans*.

---

first larval stage, and adulthood. The ventral nerve cord develops largely embryonically, with axons organized into the left and right fascicles on either side of a midline; the midline is initially constituted of motor neuron soma that are aligned along the *a/p*-axis (Boulin *et al.*, 2006) and postembryonically becomes further elaborated by an evagination ("hypodermal ridge") of epidermal tissue. The ventral nerve cord is ensheathed by basement membrane material (shown in blue). This separate organization of the fascicles of the ventral nerve cord is maintained throughout life, despite the addition of neurons and axons during the first larval stage (in yellow). However, this arrangement can be disturbed, in the absence of dedicated maintenance factors, as a result of the mechanical strain exerted by the locomotion movements of the worm (E); the failure of maintaining ventral nerve cord architecture occurs at a time right after birth (first larval stage) when additional axons and neurons are added and when the hypodermal ridge—an insurmountable obstacle in adult animals—has not yet been fully elaborated. (E) Example of maintenance defects in the ventral nerve cord. In wild-type worms, the axons of the two PVQ neurons project into the left and the right fascicles of the ventral nerve cord, during embryogenesis. These axons remain in their precise position within the ventral nerve cord throughout life, despite incessant movements of locomotion. In *zig-4*, *egl-15*, *sax-7*, and *dig-1* mutants, these axons develop normally; however, during the first larval stage, mechanical strain from locomotion leads specific axons to flip over the ventral midline to the other fascicle of the ventral nerve cord. Small white dots are background autofluorescence from the gut.